ALSO BY WENDY CHRISTENSEN

Empire of Ancient Egypt (*Great Empires of the Past* series)
The Humane Society of the United States Complete Guide to Cat Care by Wendy
    Christensen
Why Cats Do That: A Collection of Curious Kitty Quirks by Karen Anderson,
    illustrated by Wendy Christensen
Daddy Day, Daughter Day by Larry King and Chaia King, illustrated by Wendy
    Christensen

Outsmarting Cats

How to Persuade the Felines in Your Life to Do What You Want

Cats

Second Edition

Wendy Christensen

Originally published in 2004 under the title *Outwitting Cats*, a series concept created by Bill Adler Jr.

LYONS PRESS

Guilford, Connecticut

An imprint of Globe Pequot Press

With love and purrrrrrs to Jeff,
who makes everything possible.

To buy books in quantity for corporate use
or incentives, call **(800) 962-0973**
or e-mail **premiums@GlobePequot.com.**

Lyons Press is an imprint of Globe Pequot Press.

Editor: Meredith Dias
Text design and layout: Mary Ballachino

Library of Congress Cataloging-in-Publication Data is available on file.

ISBN 978-0-7627-8279-6

Printed in the United States of America

10 9 8 7 6 5 4 3 2

Contacting the author
Cat problems? Cat questions? Contact Wendy Christensen and ask:
www.wendycats.com
http://wendycats.etsy.com (Wendy's handcrafted cat-themed jewelry)
E-mail: wendy@wendycats.com
Follow Wendy on Twitter: @CatTweats

Contents

Chapter 1

Introduction

And the Cat Came Back . . .

In 1985, a quiet revolution took place: Cats overtook dogs as America's favorite companion animal. Fifty million pet cats extended their collective whiskers and edged ahead of forty-nine million dogs. The following year, felines extended their lead, and the gap has been widening ever since. According to the 2011/2012 National Pet Owners Survey, conducted by the American Pet Products Manufacturing Association (APPMA), over 86.4 million pampered, purring felines are now warming laps, snoozing in personalized kitty beds, and honing their claws on custom climbing trees in US homes. Today, 38.9 million US households are proud to call themselves cat homes.

This takeover was sweet, but it was a long time in coming. Though the feline family is ancient, humans and cats share a relatively brief history together. Compared to our long, close associations with domestic and companion animals, such as cattle and dogs, the twists and turns of the human–cat relationship have occupied only a few thousand years—the merest blink of an eye on the time scale of evolution. It's only in the past 150 years or so that we've begun tinkering with feline evolution through deliberate selective breeding.

Cats were worshipped as gods in ancient Egypt; condemned as accomplices of the devil and slaughtered by the millions in medieval Europe; treasured by seafarers as lucky charms; banished to barns, granaries, and urban jungles to fend for themselves; beribboned and elevated with women and children on a Victorian pedestal of domestic bliss. And now they've conquered the United States.

Through it all, the allure of the cat has persisted. In our busy, urbanized lives, increasingly cut off from the mysteries and rhythms of nature, we crave that thrilling aura of wildness that cats bring to our lives. We call them "domestic cats," but it's far from clear that house cats are truly domesticated. In many ways, they're indistinguishable from their closest wild feline relative, the African wildcat (*Felis silvestris libyca*), and the other small wildcats of the world. Some biologists claim

1

that our domestic cats and their small wild counterparts in Africa are really members of the same species. "The term 'domestic cat,'" notes pundit George Will, "is an oxymoron."

Trouble in Paradise

As we've taken the cat into our hearts and homes, certain . . . well, conflicts have arisen. That's not too surprising, given the inherent behavior patterns and characteristics that *Felis silvestris domesticus* brings into our homes from his life in the wild:

- Relatively small (the average cat weighs eight to twelve pounds)

- Quadrupedal (four-legged) body structure

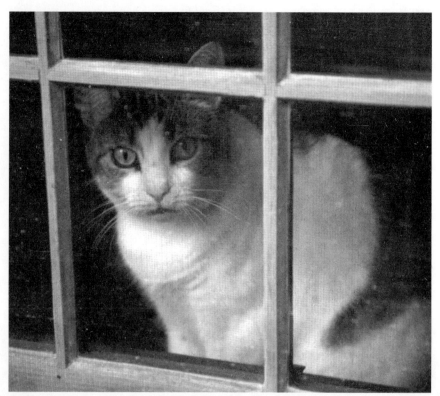

Bunny is not sure whether she'd rather be indoors or out.

- Solitary nature

- Highly territorial

- High degree of self-sufficiency

- Incredible strength for body size

- Ultrasonic hearing

- Highly sensitive sense of smell

- Additional "smell-taste" sense (the vomeronasal organ)

- Scent-oriented communication strategy

- Night-capable, motion-sensitive vision

- Sharp, powerful, retractile claws

- Teeth designed as defensive and offensive weapons as well as tools

- Hyperalertness

- Built-in wariness about being touched

Exquisitely designed for life in the wild as a solitary, self-sufficient predator, the domestic cat is less well-adapted to life in the confines of our homes, subject to human needs, preferences, prejudices, and whims. To outsmart your cat, it's important to remember that he has the same heightened sensitivity to the sights, sounds, and odors of your home as his wild cousins have to their habitats. So he's often going to act, and react, like the wild animal he still is, in so many ways.

Though he's pampered, well fed, and provided with everything he needs, what's important and meaningful to your house cat are pretty much the same things that are crucial to a bobcat prowling the snowy woods.

Can Cats Really Be Outsmarted?

Not only *can* you outsmart your cat, it's something we humans are particularly good at. Outsmarting a cat means unraveling mysteries, solving riddles, decoding cryptograms, assembling puzzles, seeing the big picture, and detecting and analyzing patterns—all skills at which the human brain excels.

Outsmarting your cat means understanding the essential truth of feline nature: Cats are solitary, independent, territorial hunters. They're superbly well-engineered and successful predators, enjoying the luxury of living at or near the top of the food chain. You'd think they could relax. But, no. Supremely self-interested, they're intensely concerned about resources: *Will there be enough for me?* Outsmarting your cat means taking advantage of his obsession with getting his share.

Can cats really be outsmarted? Silver offers his opinion.

Outsmarting your cat means learning to look at the world through feline eyes. It means realizing that your cat's behavior, especially behavior that seems naughty, perverse, spiteful, or just plain weird, makes perfect sense—to your cat.

Outsmarting your cat means learning to accept his nature, and his behavior, with good grace—or being prepared to offer him a better deal.

DISCOVER WHAT'S IMPORTANT TO YOUR CAT

To remain physically and psychologically healthy, your cat needs food, water, shelter and warmth, sufficient space, attention (on his terms) from those he cares for, opportunities to exercise his body and mind (especially through hunting), and escapability—the ability to readily and independently remove himself from situations that displease or frighten him.

In the wild and in our homes, cats live harmoniously in groups as long as each cat feels secure in his continued possession of his fair share (or more) of these resources. Feral cats live in colonies as large as available food supplies allow. If the cats are removed but the resources remain, a new group of feral cats will soon appear, again stabilizing at the ideal group size. Most wildcats aren't so fortunate. They work hard to survive and thrive, patrolling, defending, and hunting vast tracts of land. A change in their circumstances, or the arrival of a competing cat, could mean loss of territory or resources—and starvation.

Outsmarting your cat means learning how sensitive he is to change. *Any* change in a cat's familiar, comfortable environment could, in his mind, mean potential loss of resources. A change so minor or inconsequential that you don't even notice it can cause enormous stress for your cat. Many cat owners are mystified and heartbroken when changes that seem pleasant to them—new carpeting or furniture, a handsome new boyfriend, a lovely new home—throw their cats into flurries of fighting, spraying, and sulking.

Outsmarting your cat means understanding that he's a high-strung, sensitive, finely tuned creature, extraordinarily susceptible to stress and hyperalert to potential competition. A stray cat crossing the yard can throw an indoor feline into a frenzy of fear and worry. *Is that intruder going to steal my food?* It may sound silly, but to your cat, it's deadly serious. Your cat's attention is focused on feline, not human, concerns. He sees another member of his own species as a potential competitor—a threat to his resources.

Outsmarting your cat means exploiting your knowledge of what makes him tick and what's important to him to persuade him that what *you* want is also what *he* wants. And it means understanding that what's important to him is not necessarily what's important to you.

Even Up the Odds

> "In nine lifetimes, you'll never know as much about your cat as your cat knows about you."
>
> —*Michael Zullo*

Your cat already knows how to outsmart you—chances are, he does it all the time. He's the ultimate observer. He recognizes the sound of your car's engine (as distinguished from other vehicles) and the sounds your footfalls make when you're leaving, arriving, in a hurry, or at leisure. He knows your daily schedules, when you like to wake up, and when you're likely to fall asleep. He keeps a detailed, constantly updated mental notebook, registering everything that's impor-

Silver already knows how to outsmart me.

tant or interesting in his environment, including you—especially you.

Practice ESP: Everyday Strategic Planning

The intelligent, adaptable cat makes up the rules as he goes along, to suit himself
and the moment. When in doubt, adjust your own behavior, attitudes, and expectations. It'll be a lot easier than adjusting your cat.

> "No tame animal has lost less of its native dignity or maintained
> more of its ancient reserve. The domestic cat might rebel tomorrow."
> —*William Conway, archbishop of Armagh*

And you'd better be ready!

Outsmarting your cat (or cats) is an everyday activity, not just an emergency
response to a crisis. To outsmart your cat, you need to stay at least a step ahead of
him all the time. Note well his concerns, his behavioral quirks, what worries him,
what frightens him, what his various meows mean, what pleases him, and what
tends to unhinge him. In discovering and noting the collection of behaviors, attitudes, and quirks that make each cat unique in all the world, you're building up a
priceless strategic resource: your own mental notebook. With this, you're ready
to follow the Boy Scouts' advice: Be Prepared!

> "Always the cat remains a little beyond the limits we try to set for
> him in our blind folly."
> —*Andre Norton*

Nurture a Well-Tended Relationship

The best way to be confident that you'll be in a position of strength when you
really need to outsmart your cat is by tending your relationship every day. This
won't prevent all cat problems, but it'll help set you up for success. A well-tended
relationship lets you:

1. Detect problems when they're still small—before they mutate into huge,
   full-blown crises (which can happen quicker than you think!).

2. Maintain a built-in strategic advantage in solving problems that do
   arise.

3. Build up your major strategic weapon: your mental notebook of observations about the peculiarities and idiosyncrasies of each cat in your life.

4. Reduce stress on both you and your cats when problems do arise, because your knowledge gives you more options.

What's My Name?

Your cat knows his name quite well. He may choose to respond to it sometimes and ignore it other times. It's his option or whim. He also may disapprove of the name you've given him and prefer another. How can you find out what this is? Just ask.

If you ask, over time, your cat will coach you to discover the name he'd prefer to be called. Watch him carefully, listen to your intuition, and try out different names. See how he responds. When you hit on the right one, you'll both know.

Sustain a Lifelong Conversation

Your cat is amazingly adept at reading and interpreting your moods, tones of voice, emotions, habits, and routines. He knows instantly if you're unhappy about something he's doing. (He doesn't always care—but he *does* know.) I talk to my cats all the time, and they talk back, each in his or her own dialect of "catspeak." We share a lifelong, running conversation.

While my cats are the experts in cat matters, they seem to consider me an authority on the larger world. If there's a sudden loud noise or rambunctious visitors, the crackling and booming of a thunderstorm, or the *beep-beep-beep* of a flying saucer in the front meadow—my cats look to me. If I stay calm, they usually do, too. Veterinary ethologist (expert on animal behavior) and author Myrna Milani, DVM, says it best: "Calm owner, calm cat."

I can't emphasize this enough: Calm owner, calm cat. Calm owner, calm cat. Make it your mantra.

NEVER INSIST

"Dogs live to please their owners. Cats live to please themselves."
—*Phil Maggitti*

Cats choose to interact with us at their own pace, on their own terms. Your cat will be much happier and more contented if you refrain from forcing or insisting upon

Mewphy's Laws™

It's time cat owners learn what engineers and scientists have long known. Legend has it that back in the 1940s, a certain Captain Murphy, after observing how often, persistently, and thoroughly things went wrong in his military testing laboratory, codified his famous "Murphy's Laws," the first of which is the universally true: "Anything that can go wrong will go wrong."

Well, it seems Captain Murphy had a cat, Mewphy. And while Murphy was codifying the laws of the human universe, Mewphy was busy discovering the laws that govern cat problems. Now, for the first time, Mewphy reveals his "Universal Laws of Cat Behavior and Cat Problems."

1. Left to themselves, most cat problems go from bad to worse.

2. Left to themselves, some cat problems get better, and meddling or interfering actually makes them worse.

3. Learning the difference between #1 and #2 is often a lifetime project, especially without help. (Aren't you glad you found *Outsmarting Cats*?)

4. Cat problems that call for the least attention and intervention (such as intercat conflicts) usually get the most, while cat problems that call for the most intervention (such as litter box problems) generally get the least.

5. If you think you smell cat urine, you probably do.

6. Cat urine, left to itself, multiplies. ("Urine attracts urine.")

7. Yes, it's soaked into the padding. And the floorboards. And maybe the floor joists, too (the "law of anti-carpetarianism").

8. "'Tis better to ask for forgiveness than permission."

any particular style or amount of daily interaction. *Never insist.* This can take some psychological adjustment, especially for people more accustomed to living with dogs. Dogs are more like palace courtiers—they dance attendance upon us, await our pleasures, live to serve.

Cats do no such thing.

And expecting them to is a recipe for disaster—or at least mutual misunderstanding and unhappiness.

9. Self-cleaning litter boxes aren't.

10. An empty litter box is not a good sign.

11. If you pester a cat long enough, you will get scratched.

12. You will deserve it.

13. If you live with a group of cats and nothing seems to be going wrong, you have probably overlooked something.

14. It is impossible to make anything truly cat-proof because cats are so ingenious.

15. It is in the nature of cats to hide problems. What you don't see is as important as what you do see.

16. The most interesting place in the world is on the other side of that closed door.

17. As soon as the door is opened, #16 ceases to be true.

18. It's much easier to change your environment than change your cat.

19. It's much easier to adjust your attitude than adjust your cat.

20. "Everything here is mine" (a message from Mewphy—and your cat).

> "There are people, who reshape the world by force or argument, but the cat just lies there dozing; and the world quietly reshapes itself to suit his comfort and convenience."
> —*Allen and Ivy Dodd*

Living with a cat calls for thoughtful accommodation to feline physical and psychological needs and natures, and an appreciation of the adaptations and sacrifices he's making to fit as smoothly as he does into your life. A calm, philosophical acceptance of a certain amount of damage to your possessions over time is part of the equation of living in the close company of cats. On balance, the payoffs are well worth the small headaches and frustrations. Living with these thrillingly semi-wild creatures is both a challenge and a joy.

Cats have their own society, their own customs, their own agendas.

Enjoy the Adventure

Considering how close domestic cats are to their wild counterparts in form and function, behavior, and outlook, it's amazing how well they thrive in our homes and our world. Their degree of adaptation to our quirks, ways, and lifestyles is a tribute to their intelligence, tolerance, adaptability—and to their thoroughgoing opportunism.

The best way to outsmart your cat is to enjoy the adventure together. Let him be a cat. Provide a safe, stable, peaceable, enriched environment in which he can become the best cat he can be. Let him indulge in the full range of normal, natural feline behavior, in ways compatible with your own needs and home environment:

- Give him space—to roam, play, hide, wander, ponder, investigate.

- Look up. Extend his range to three dimensions.

- Learn the subtleties of vocalized catspeak, and pay attention to feline body language. Both are full of clues to your cat's mood and intentions.

- Offer daily opportunities for predatory mock-hunting play.

- Provide cat-attractive scratching sites.

- Accommodate your cat's craving for private dens and hideaways.

- Offer foraging opportunities—great physical and mental exercise.

- Be creative. Devise cat-oriented problem-solving exercises.

- Include your cat in your daily routines and activities.

The Fivefold Path to Human–Feline Fulfillment

1. Ignore feline behavior you *don't* like.

2. Reward and praise feline behavior you *do* like.

3. Channel your cat's energies in acceptable directions.

4. Don't place undue temptation in your cat's path.

5. Be consistent.

Chapter 2

People and Cats, Together

In the Beginning . . .

The human and feline races are millions of years old, but it's only in the last few thousand years that people and cats have kept company. Small cats were settled into broad swaths of Africa and Asia for millennia before our wandering ancestors took notice of them. It should come as no surprise that when cats and humans came together, it was most likely the cats' idea—and that cats made the first move, for reasons of their own.

Your cat's first distant ancestors, tree-dwelling miacids, appeared about sixty million to seventy million years ago, during the Paleocene epoch. Miacids, likely evolved from earlier placental mammals, are the ancestors of all modern mammalian carnivores, including felines, canines, and bears. Similar to a modern pine marten, *Miacis* had a long, athletic, weasel-like body and, possibly, retractile claws.

The relatively sudden end of the age of dinosaurs, about sixty-five million years ago, changed everything. Without threat or competition from these long-dominant creatures, the stage was set for the swift rise of the most adaptable, cleverest, and best-equipped mammals. *Miacis* had a larger brain than other early carnivores, giving her a critical edge in the evolutionary struggle for survival and dominance.

She also had remarkable teeth that were, like your cat's, specialized for cutting and tearing meat. In both the upper and the lower jaw were set incisors, canines, premolars, molars, and carnassials. The carnassials, with their scissor-like action for shearing meat into chunks, were a critical evolutionary innovation, giving *Miacis* a tremendous advantage over her flesh-eating competitors. *Miacis* evolved rapidly, leaving her challengers in the dust of extinction.

Her descendants developed specialized forms and behaviors to exploit a wide range of ecological niches. Until humans appeared on the scene, miacids and their descendants dominated the food chain virtually everywhere in the ancient world. One of these descendants, *Profelis*, appeared about forty-five million years ago and is the ancient forebear of the lion—and your cat.

By only a few million years later, according to the fossil record, the descendants of *Profelis* had split into two lines—*Hoplophoneus* and *Dinictis*. Both had long bodies and tails, short legs, and a civet-like appearance. Though they still walked on flat feet (a *plantigrade* walk) rather than on their toes (a *digitigrade* walk) like modern cats do, they were clearly feline in silhouette and habits. *Hoplophoneus* went on to found the doomed line of overly specialized saber-toothed cats.

Smarter and more adaptable, *Dinictis* prospered. This lynx-sized carnivore had the efficient teeth of her ancestors, but more refined and better balanced between upper and lower jaw than the unfortunate saber-toothed cats. *Dinictis*, like your cat, had a "third eyelid," or nictitating membrane—a protective adaptation that let her stalk prey more safely and swiftly in dense foliage and tall grass. Though smaller-brained than modern cats, *Dinictis* was a powerful, successful predator, well adapted to numerous ecological niches. She and her descendants spread widely and rapidly, evolving by thirty million years ago into *Proailurus*, the first true member of the Felidae, the modern cat family.

Family Felidae Appears

Proailurus was blessed with keen senses, a relatively large brain, and retractile claws for gripping prey and sure-footed, speedy climbing. Though not quite tabby snoozing on your lap, *Proailurus* walked on her toes, like your cat, and her distinctive teeth, optimized for chopping and tearing meat, were just like your cat's teeth.

Although the oldest fossil remains of modern cats are only three million years old, *Proailurus* and her progeny, *Pseudoailurus*, had spun off the modern "small cats" by the late Miocene epoch, about twelve million to eighteen million years ago. By three million years ago, today's three feline genera—*Panthera* (the large or "roaring cats," such as lions, tigers, and leopards); *Acinonyx* (the cheetah); and *Felis* (the small or "purring cats")—graced the earth. The modern cheetah, the only descendant of *Acinonyx*, is little changed from her ancient ancestor, and sports the only non-retractile claws in the cat family. To the mighty *Panthera*, who once ranged much farther than they do today, our ancestors were strictly prey.

Feline family portrait.

The highly adaptable small cats diverged into numerous specialized subgroups and spread rapidly, exploiting a vast range of climate, terrain, and prey. Among the earliest small cats was Pallas' cat (*Felis manul*, now classified as *Otocolobus manul*), the oldest still-living species of the genus *Felis*.

Between 600,000 and 900,000 years ago, in an era between two great ice ages, *Felis silvestris*, a small, fierce, striped wildcat, prowled the endless forests of northern Europe until the ice returned and relentlessly pushed her south. Crossing temporary land bridges across the Mediterranean, she padded into Africa and Asia and prospered. Those few hardy representatives of *F. silvestris* who remained in northern Europe and the British Isles evolved into the elusive Scottish or European wildcat (*F. s. silvestris*).

In Asia, *Felis silvestris* developed into the Chinese desert cat (*F. bieti*). Across the vast continent of Africa, *F. silvestris* adapted to changing climate, terrain, prey, and living conditions and eventually branched into the sand cat (*F. margarita*), jungle cat (*F. chaus*), black-footed cat (*F. nigripes*), and African wildcat (*F. libyca*). Currently recognized as a subspecies and classified as *F. silvestris libyca*, the African wildcat was the foremother of today's domestic cats.

For most of human history, our ancestors were ill prepared to host the cat in the style to which she intended to become accustomed. No doubt our cats' ancestors watched our hunter-gatherer forebears from a distance, warily observing the highly inefficient hunting styles of those small bands of strange, muttering bipeds. Always on the move, following the seasons, ancient humans tracked elusive game and rooted out edible plants, fruits, nuts, and berries. It was not a lifestyle to entice an independent, self-sufficient, place-loving predator. So, in Rudyard Kipling's words, "The cat still walked alone."

The earliest known remains of small cats living among humans were discovered in 2004 by archaeologists at a burial site at Shillourokambos, a stone-age village on the Mediterranean island of Cyprus. The carefully buried eight-month-old cat (probably *Felis silvestris libyca*) lay only sixteen inches from the remains of a thirty-year-old person, beneath a layer of earth about 9,500 years old. The human grave included red ocher dye, seashells, jewelry, polished stones, axes, and flint tools. The human and cat skeletons were positioned symmetrically, with both heads pointing west, and in identical states of preservation. Based on the goods found in the grave, the person was obviously of high rank. Was the cat his pet, or did it have some symbolic significance? The careful, reverent burial of the cat seems to indicate a close bond between the two—a bond they carried into the afterlife.

Also found at the Cyprus site was a cat figurine, even older than the joint human–feline burial. Archaeologists see this as a sign that the people living there regarded cats as highly significant. But it isn't known whether the cats of Cyprus were actually domesticated.

Other remains of small felines (also likely *Felis silvestris libyca*) were unearthed at a Neolithic human settlement site at Jericho, Palestine. They date from around 6700 BCE (Before the Current Era), almost nine thousand years ago. A 3,600-year-old wall mural made of faience tiles at the Palace of Knossos on the Mediterranean island of Crete depicts hunting cats (similar to cats seen in Egyptian wall paintings) with wild sheep.

It was long thought that cats were first domesticated in ancient Egypt—a natural conclusion, based on the rich heritage of feline art the Egyptians left behind. But with the recent sequencing of the feline genome (see "The Feline Genome: Helping Cats and Humans," page 16), we now have a much clearer window into the early domestication of cats, including when and where it first happened. In 2007, scientists from the Laboratory of Genomic Diversity at the National Cancer Institute in Bethesda, Maryland, analyzed genetic material from about one thousand modern small cats, both domestic and wild, from Africa, Asia, and Europe.

The Feline Genome: Helping Cats and Humans

Cinnamon, a quiet Abyssinian with a long, well-documented pedigree, lives with researchers at the University of Missouri College of Veterinary Medicine in Columbia. As the first domestic cat to have her genome sequenced, Cinnamon is a heroine to all cats—though she'd be much too shy to brag about it.

The successful mapping of Cinnamon's genome, a landmark scientific achievement—was reported in the journal *Science* in October 2007.

This research, conducted by scientists collaborating from multiple laboratories, produced a tremendous amount of new information not just about cats, but about all mammals, including us. By studying Cinnamon, researchers were able to identify over twenty thousand feline genes, and to discover important clues about how and when cats have changed and evolved over the millennia.

Cinnamon's well-known ancestry helped scientists confirm that we and our cats really are a lot alike. Because of Cinnamon, we now know that over 90 percent of the cat's genes are derived from the same common ancestors as the comparable genes in humans.

Scientists have long known that by studying cats, they can learn a lot about people. Cats suffer from over 250 health conditions and hereditary problems, many similar or identical to human illnesses. These include some of our commonest and most feared health threats: numerous infectious diseases, diabetes, cancers, heart disease, eye problems, polycystic kidney disease,

The scientists found that the wildcats and domestic cats could be divided into five subgroups, or subspecies, of the small wildcat *Felis silvestris*: the European wildcat (*F. s. silvestris*); the southern African wildcat (*F. s. cafra*); the central Asian wildcat (*F. s. ornata*); the Chinese desert cat (*F. s. bieti*); and the Near Eastern wildcat (*F. s. libyca*). All of the domestic cats, whether pedigreed or randomly bred, fancy or feral, are descended from the Near Eastern wildcat group. *Felis silvestris libyca*, the common ancestor of all domesticated cats, lived in the Middle East and split off from the other wildcat groups about 130,000 years ago.

These great-great-grandfathers of your cat were small wild felines who lived in the area called the "Fertile Crescent," ranging from the eastern Mediterranean around Turkey and down through Mesopotamia (present-day Iraq). This is precisely the region in which our own ancestors gave up their old hunter-gatherer lifestyle in favor of farming about twelve thousand years ago. Based on the varying lineages of mitochondrial DNA the researchers found in the modern cats they studied, domestication probably occurred at least a half dozen times in the Midrespiratory problems, muscle diseases, and many more. Significantly, cats are the only animals besides people known to naturally get ill from immune deficiency viruses, such as HIV (in humans) and FIV (in cats).

The first big news from Cinnamon's genome was related to her Abyssinian pedigree. Abyssinians are more likely than other cats to suffer from *retinitis pigmentosa*, a serious eye disease that can cause blindness, which also affects millions of people around the world. Discoveries of genetic similarities in the human and feline forms of the disease have already pointed to promising new treatments for both humans and cats.

Cinnamon's genome has also led to the discovery of the genes responsible for about a dozen feline diseases (so far). Breeders of certain breeds of cats can now determine, through DNA screening, whether their breeding cats carry the genes associated with polycystic kidney disease and hypertrophic cardiomyopathy, among other diseases.

Because of the successful work with the Feline Genome, a wealth of new disease prevention strategies, and even cures, are on the horizon for cats. In 2011, the Cat Health Network, a collaboration of the Morris Animal Foundation, the Winn Feline Foundation, the American Association of Feline Practitioners, and the American Veterinary Medical Association, was launched. This new collaborative effort was begun in response to the sad fact that cats were falling behind dogs in visits to the veterinarian, and the equally sad fact that research into cat health gets far less money and attention than dog health research.

And it all began with a shy Abyssinian cat. Thank you, Cinnamon!

dle East, around eight thousand years ago. All of today's domestic cats (*Felis silvestris catus*)—from showiest pedigreed beauty to humblest feral—are descended primarily from at least five *Felis silvestris libyca* "founder" cats. Genetic and anatomical research also shows some later interbreeding with *F. s. silvestris* and *F. s. ornata* along the way.

A Seed Is Planted

Between nine thousand and twelve thousand years ago, a shift from food gathering to food production gradually took place in the Near East (now called the Middle East). The region between the Tigris and Euphrates Rivers offered fertile soils and sufficient rains to nourish wild barley and wheat for humans, and grass for domesticated pigs, goats, and sheep. Nine thousand years ago, the people there had domesticated these grains and animals and were living in villages near their herds and fields.

Agriculture—the "Neolithic Revolution"—spread rapidly beyond the Fertile Crescent. By seven thousand years ago, it had spread to the Balkan Peninsula; by six thousand years ago to Egypt and central Europe. By five thousand years ago humans ranging from Britain to northwest India were settling down, tilling fields, harvesting crops, and storing their produce against need. Rodents took notice.

Storehouses and granaries, where farmers stored the hard-won produce of their labors, had been quickly colonized by hordes of rats and mice, which consumed huge amounts of precious grain and fouled many times more. The most curious and adventurous of the local wild felines wandered into farming villages, finding the hunting splendid and the company congenial. Cat heaven! Gradually, *Felis silvestris libyca* settled in. As humans moved about the region, whether over land or on ships, cats came along, too.

The Cat Comes Home

One of the places the cat found herself at home was a narrow valley in northeast Africa, where a mighty agricultural civilization had prospered from the Nile River's annual life-giving flood. This small, long-legged, sandy-colored tabby, sometimes called "the gloved cat" for her distinctive forearm markings, was a godsend, and later a goddess, to the ancient Egyptians.

The newcomers efficiently protected the valuable fruits of the harvest, while charming one and all with their beauty, grace, cleanliness, independence, and aura of mystery. These twin aspects of gentle companionship and fierce protectiveness appealed to the Egyptians' religious and artistic sensibilities. It's no wonder they elevated the cat into their pantheon in the form of Bast, or Bastet, a goddess with the body of a woman and the head of a cat. Bast was goddess of all good things: wisdom, music, dancing, fertility, sensual pleasure, happiness, warmth, and basking in the sun. Goddess of the moon, Bast

In ancient Egypt, cats were worshipped. They have not forgotten this.

The Cat in Ancient Egypt:
Godsend and Goddess

Thousands of years ago, the ancient Egyptians created exquisite artistic representations of felines that have never been equaled. Most of the cats depicted in Egyptian art are of the species *Felis silvestris libyca,* the small African wildcat who is the foremother of all our domestic cats. Egyptian artists depicted cats both as tame domestic companions and as fierce protectors and hunters.

By 4,500 years ago, cats were cherished companions in Egyptian homes, as well as tireless and efficient protectors of the food supply. The first Egyptian depiction of a cat with a collar appears in a tomb at Saqqara dating from the fifth dynasty (c. 2500–2350 BCE). In a Theban tomb painting, a ginger tabby plays beneath his owner's chair, wearing a collar with a lead tied to a chair leg. Another painting shows an Egyptian gentleman with one kitten perched on his lap and another, adorned with a silver earring, posing prettily beneath his wife's chair. In one famous wall painting from Thebes (c. 1450 BCE), a large tabby cat joins her owners on a bird-hunting expedition to the Nile marshes. Clutching one large bird in her teeth, she springs from the hunter's boat to grab two others with her claws!

Cats adorned jewelry, figurines, and musical instruments. Women ringed their eyes with kohl (a type of eye makeup) in imitation of their cats' distinctive tabby markings.

A popular name for little girls was Miit, or "cat." One court lady was nicknamed Ta-miit ("She-cat"). An Egyptian king was called Pa-miu ("The Tomcat"). Greek writer Diodorus observed that the Egyptians called their cats with a "special clucking sound" and fed them bread in milk as well as cut-up fish.

Although most ancient Egyptian felines remain eternally nameless, records of a few early feline names have survived. The earliest is that of the royal cat Nedjem, which means "pleasant" or "sweet." Nedjem was, like most cats of ancient Egypt, a long-legged "gloved cat," or mackerel tabby. Mackerel tabbies, with their "fishbone" stripes, were much rarer, and thus more highly prized by ancient Egyptian royalty, than what we now call "classic" tabbies (tabbies with bull's-eye markings). Nedjem was a pampered royal pet, part of the court of the Pharaoh Thutmose III (1479–1425 BCE, known as the "Napoleon of Egypt"). Another royal pet feline, Ta-miaut (also referred to as Tamyt or "The She-Cat"), was the cherished companion of Crown Prince Thutmose, the eldest son of Pharaoh Amenhotep III (1386–1349 BCE). Ta-miaut was mummified and buried in a beautifully decorated limestone sarcophagus. In an inscription on her sarcophagus, Ta-miaut proclaims to the sky goddess, Nut, that she—like all felines!—is a star: "I myself am placed among the imperishable ones that are in the Sky, for I am Ta-Miaut, the Triumphant."

How wonderful it must have seemed to the Egyptians to enjoy the companionship of these graceful, amusing, self-cleaning creatures who also protected their precious food supplies. What light it must have brought to their eyes to delight in the snuggles and purrs of their small, elegant saviors.

Cat Proverbs

In the many lands they settled, cats have inspired folklore, wisdom, and proverbs.

Ancient Egypt: "Thou art the Great Cat, the avenger of the Gods, and the judge of words, and the president of the sovereign chiefs and the governor of the holy Circle; thou art indeed . . . the Great Cat."
—from an inscription on the Royal Tombs at Thebes

America: "You will always be lucky if you know how to make friends with strange cats."

Arabia: "The cat was created when the lion sneezed."
"A cat bitten once by a snake dreads even rope."

Britain: "In a cat's eye, all things belong to cats."
"A cat may look at a king."
"A cat has nine lives. For three he plays, for three he strays, and for the last three he stays."
"Curiosity killed the cat, satisfaction brought it back!"
"The dog for the man, the cat for the woman."

China: "Happy owner, happy cat. Indifferent owner, reclusive cat."
"I gave an order to a cat, and the cat gave it to its tail."
"When rats infest the Palace a lame cat is better than the swiftest horse."

held the fire of the sun in her eyes overnight, preserving its light and warmth for her people. How appropriate that the ancient Egyptian word for "cat" was *miu* or *mau*—which also means "light."

Seagoing Saviors

Even before the invention of agriculture, adventurous humans explored the Mediterranean Sea and beyond in wooden ships. Rodents were a constant menace aboard ships, devouring and fouling cargoes and food supplies. Curiosity—and a nose for easy, concentrated prey—likely enticed the small, adventurous, opportunistic native cats of northern African ports to trade their dockside homes for sailors' berths. Once at sea, the stowaways grew sea legs and made themselves at home in their new "territories." They quickly proved their worth by dispatching rats and mice, while providing amusement and companionship for sailors.

Ethiopia: "A cat may go to a monastery, but she still remains a cat."

France: "Books and cats and fair-haired little girls make the best furnishing for a room."
"Cats, flies, and women are ever at their toilets."
"The cat is nature's Beauty."

Germany: "The cat who frightens the mice away is as good as the cat who eats them."

Italy: "Happy is the home with at least one cat."

Ireland: "Beware of people who dislike cats."

Morocco: "An old cat will not learn how to dance."

Native American: "After dark all cats are leopards."

Wales: "Ye shall not possess any beast, my dear sisters, except only a cat."
—The Ancren Riewle (Nun's Rule)

Cats quickly graduated from utilitarian ratters to the cherished "guardian spirits" and good-luck charms of their chosen vessels. If something happened to the ship's cat, the voyage was considered doomed. The ancient tie between cats and seafarers resonated well into modern times: Until 1976, the British navy required ship's cats on all vessels.

The ancient Egyptians so revered their cats that they strictly prohibited feline exports. But confining such a free spirit—not to mention such a useful commodity in an increasingly agricultural world—proved impossible. By 900 BCE, seafaring traders and smugglers were already carrying domestic cats all over the Mediterranean and beyond. These tabby-striped, short-haired felines, prized wherever agriculture was practiced and rodents threatened (nearly everywhere), spread rapidly throughout Europe and Asia.

Today, descendants of the first seafaring felines are found on every continent (including Antarctica, under human protection). Almost all of them got there originally aboard ships.

Chapter 3

The Physical Cat

In outsmarting your cat, knowledge is power. Knowing how your cat's built, and understanding how his physical systems function individually and together to support his natural role as a successful predator, will help you better understand and appreciate his mind, motivations, and way of seeing the world.

Your Cat's Skeleton

Your cat is supported by a light, sturdy framework. Like all felines, your cat is a perfect predator from his skeleton up: powerful, agile, stealthy, and exquisitely controlled. Consisting of about 244 bones (40 more than humans), your cat's inner scaffolding, working in concert with hundreds of skeletal muscles in a responsive system of levers, endows him with agility, speed, power, suppleness, and extraordinary grace.

Silver demonstrates his amazing flexibility.

The central feature of the feline skeleton is the *spine*, or backbone. More than thirty-three different-shaped *vertebrae* (not counting tail vertebrae) are interleaved with cartilaginous *disks* that absorb shocks and provide stunning maneuverability. The vertebrae are connected more loosely than in humans and other animals, allowing cats to curl into tight balls, roll over with a half twist on their backs—or even in midair—and stretch out to what seems like impossible lengths. When your cat gallops at full speed, his spine alternately flexes and stretches to enhance forward momentum.

The skull, optimized for a stealthy hunter's needs, features huge eye sockets; a powerful jaw with well-developed anchor points for muscles that aid the killing bite; and an array of specialized teeth. Where the skull joins the spine, the *cervical vertebrae* form your cat's neck. The tiny *hyoid bone* located in the neck prevents him from roaring like the big cats, but allows a continuous purr—not a bad trade-off.

As your cat stalks, his *scapulae*, or shoulder blades, rise and fall while his head and shoulders remain virtually motionless. Shoulders swinging along with his legs, your cat enjoys both a lengthened stride and a greater range of motion. The scapulae also anchor his *forelimbs*.

Delicate-looking but sturdy, the bones of the forelimbs provide stability and maneuverability. The *wrist joint* is extraordinarily flexible, twisting to allow him to use his paws and claws for grasping prey, swiping at foes, climbing and scrambling, and self-grooming.

The long, slender, curved bones of the rib cage, anchored along the *thoracic vertebrae*, support and protect vital organs such as the heart and lungs. An unusual feature of the feline skeleton is the floating *clavicle*, or collarbone. Not attached to the shoulder joint, as in humans, the clavicle "floats" within muscle, allowing your cat to squeeze through tiny openings and thread narrow passages with speed and precision by placing one forelimb directly ahead of the other.

Behind the *lumbar vertebrae* of the lower back, the relatively thick *pelvic bone* (or *pelvic girdle*) anchors the powerful hip joints. The strong *hind limbs* offer power, push, and spring. Though not quite as mobile as his forelimbs, your cat's rear paws are formidable weapons and useful tools.

At birth, a kitten's bones are not fully *ossified*, or hardened. The kitten's pliable, cartilaginous skeleton enables his bouncy, gravity-defying antics. As kittens bloom into cats, bones harden and strengthen, but cats never lose their agility, flexibility, and grace.

"A kitten is so flexible that she is almost double; the hind parts are equivalent to another kitten with which the forepart plays. She does not discover that her tail belongs to her until you tread on it."

—*Henry David Thoreau*

Your Cat's Tail

Your cat's versatile tail—handy for both balance and communication—is an integral part of his spine, accounting for a third of its length. The tail contains from just a few to over two dozen specialized bones called *caudal vertebrae*, and joins the body at the *sacral vertebrae. Sacrocaudalis* muscles enfolding the caudal vertebrae help give the tail its ability to maintain balance and to communicate your cat's mood and state of mind.

Like a tightly controlled string of beads, the caudal vertebrae can instantaneously twist, curl, stiffen, and compress to help balance and stabilize a darting, leaping, or pouncing cat. As he gallops full speed, your cat's tail serves as a rudder, a counterweight to compensate for course corrections. The tail's balance-beam ability helps your cat accurately and swiftly shift his center of gravity as he navigates narrow fence tops without faltering or breaking stride.

A kitten's balancing skills are perfected early. His constant active play hones his pouncing, leaping, and stalking skills, and helps him build confidence in his power, quickness, and flexibility.

No injury to your cat's tail is trivial. All require immediate veterinary attention. Fractures can be serious. A traumatic blow to a cat's hind end can separate the first caudal vertebra from the sacrum at the tail head (where the tail is attached to the body), permanently paralyzing the tail. Fractures and dislocations nearer the tail tip usually require amputation. Besides pain, tail injuries can also cause life-threatening bladder paralysis. A cat who loses his tail to trauma may take weeks or months to regain even a semblance of his former confidence and agility. Some tail-injured cats never recover their sense of balance.

Your Cat's Teeth

Your cat doesn't chew his food. He chops it up and swallows the chunks. Unlike the teeth of omnivores (like us) and herbivores, your cat's teeth don't have flat upper surfaces for chewing and grinding vegetation. Instead, they're optimized for stabbing, chopping, shearing, and stripping meat from bone.

Your cat's versatile teeth also come in handy for defense and offense. His canines, or fangs, are designed for dispatching prey swiftly and accurately with the trademark "killing bite."

Kittens are born toothless. By two months, they have twenty-six deciduous baby teeth. By five to six months, thirty permanent

Jessie shows off her splendid dentition.

teeth replace them. Your cat's six *upper incisors* and six *lower incisors* are used for gentle nibbling. He uses his four *canine* teeth ("fangs") to stab prey and adminis-ter the killing bite. His *upper premolars* (six in all, three per side) shear and grind food, while his two *upper molars* (one on each side) chop and cut.

His four *lower premolars* (two per side) grind, while his two *lower molars* (one on each side) shear meat from bone. Your cat's lower molars and last upper premolars (called, together, the *carnassials*) fit together like scissors, enabling your cat to chop his food into chunks before swallowing.

Your Cat's Muscles

Your cat has more than five hundred *skeletal muscles*; we much larger humans have only about six hundred.

Although he also has muscles not under voluntary control (including *cardiac muscle* and *smooth muscles* in internal organs), it's his skeletal muscles that endow your cat with his stunning speed and power. Large, strong muscles attach directly to his pelvic girdle. While galloping all-out, his agile forelimbs bear almost half his weight. His hind limbs become mighty engines of forward propulsion.

Your cat's floating collarbone and narrow chest allow his forelimbs to swing freely; the forelimbs and paws contribute flexibility and agility. His *digitigrade* ("on toes") posture offers both speed and stability, effectively lengthening his limbs and thus his overall stride. Only a small portion of what in humans is called the ball of the foot ever touches the ground, allowing the limb to quickly move forward, accelerating movement at any speed.

Your cat is a "perfect walker," placing each hind foot almost exactly in the print made by the corresponding front foot, making a nearly straight-line track. The ultimate conservationists, cats expend no more energy than is absolutely necessary. The typical cat gait is a leisurely walk. Although your cat has trouble changing direction rapidly (you've seen him skidding to a halt and fishtailing around corners), he's a master at shifting swiftly from one gait to another.

Like only two other animals, the camel and the giraffe, all felines share a "diagonal" gait (both legs on one side, then both on the other) that requires a minimum of energy while optimizing grace, agility, stealth, and speed. Starting out at a walk, your cat first extends his right hind leg, then his right foreleg, then his left hind leg, and finally his left foreleg. If an interesting opportunity motivates a brisker trot, his hind legs move more rapidly to catch up with the diagonally opposing forelegs, and then move in pairs: right foreleg with left hind leg, left foreleg with right hind leg.

In a heartbeat, the trot explodes into a mighty gallop, a series of long, low half bounds, his flexible spine alternately stretching and compressing. Picking up speed, your cat powers himself forward from both hind limbs at once, lands ever so lightly on one forefoot, and immediately transfers his weight to the other forefoot, which touches down a stride farther on. His hind limbs touch down again, propelling him forward with their mighty thrust. At full throttle (he can sprint about thirty-one miles per hour), his hind limbs can extend ahead of his forelimbs, and he's actually airborne between strides. He can sustain this speed only for short bursts, though. His sprinter's muscles tire quickly.

Whether pursuing prey or escaping danger, cats prefer jumping over sprinting. Your cat can easily jump five times his height. When it comes to climbing, what goes up (easily and quickly) must come down (not so easily). Superbly designed for ratcheting upward with powerful thrusts from his hind limbs and a secure grip with his foreclaws, your cat has considerably more trouble reversing the process. He'll generally resort to slithering down, tail-first, until he's within striking distance of the ground, and then turn and jump.

Every bit as vital to your cat as swift, efficient movement is his ability to instantly stop and freeze, holding a pose impossibly long. His musculature also endows him with the capacity for extremely slow, highly controlled stealthy movement—that maddening, barely perceptible creep that suddenly explodes into a mighty pounce.

Your Cat's Skin and Coat

Your cat's skin is his largest organ. It's thinner than that of most other animals: ¹⁄₁₆ to ¹⁄₁₂ inch thick over his neck, but only ¹⁄₆₄ inch thick on his belly. Feline skin is a complex, multifunctional system, rich with receptors sensitive to touch, temperature, pressure, vibration, and pain. It regulates heat loss, keeping vital moisture in and toxins and harmful bacteria out. Its looseness ensures that most wounds remain superficial, and it can even aid your cat's escape from an enemy's grasp.

The inner layer, or *dermis*, bristles with sensors and complex structures, especially in the *follicles*, pits in the skin from which hair grows. As the cells of the skin's outer layer, or *epidermis*, multiply and migrate outward, replacing dead cells shed at the surface, they flatten and become horny, forming a tough, waterproof barrier. *Pigment* in the epidermis gives the skin its color and protects against harmful solar radiation. Tiny bumps called *tactile pads*, on the skin between the follicles, respond to the slightest pressure.

His hair coat is your cat's first line of defense, providing insulation, waterproofing, camouflage, and protection against minor cuts and abrasions. Your cat's hair (up to 130,000 hairs per square inch on his belly, and about half that along his back) grows about ¹⁄₁₂ of an inch every month. Each *primary* or *guard hair* grows from an individual follicle, surrounded by clusters of *secondary hairs*—thickened, bristle-tipped *awn hairs*, or fine, crinkled *down* or *wool hairs*. Secondary hair grows in clusters around each guard hair.

Each guard hair's follicle has its own blood supply, nerves, touch receptors, and *sebaceous gland*. *Sebum* secreted by this gland gives each hair a waterproof shield and endows the coat with its healthy shine. Sebum also contains *cholesterol*, converted by sunlight to essential vitamin D and ingested when your cat grooms himself. The *arrector muscle* connected to each follicle lets your cat instantly fluff out his coat in response to cold (to trap warm air near his skin) or to a threat (to make him appear larger and more formidable).

Each follicle has its own growth cycle, a period of rapid hair growth followed by a slowdown and then a resting phase. When the follicle becomes active again, new growth pushes the old hair out—right onto your sofa or suit. See chapter 11 for hints on outsmarting your cat's predilection for shedding all over your furniture and clothing.

Your cat's tough, protective, hairless foot pads are up to 75 percent thicker than the skin on any other part of his body, providing silent, sure-footed tread and

Are Black Cats Lucky Cats?

According to scientists at the US National Institutes of Health, black fur has evolved separately many different times in different species of cats, including the domestic cat, the leopard, and the jaguar. The fact that a particular mutation shows up repeatedly usually means that the mutation gives the animal some survival advantage. Of course, black fur itself is a great survival advantage: It makes stalking and hunting in the dark much easier!

Owners of black cats now have an extra reason to prize their lucky felines. Modern genetic research has revealed that black-furred cats may be more resistant to certain kinds of diseases than other cats. Evolutionary geneticists Eduardo Eizirik and Stephen O'Brien and their colleagues, working at the NIH's National Cancer Institute in Bethesda, Maryland, found that a mutation of a gene called MC1R causes black fur in cats. Interestingly, MC1R is part of a group of genes, the 7-transmembrane receptors, that allow viruses such as HIV to enter cells. Sure enough, black leopards (also called "melanistic leopards") proved to be better at surviving virus epidemics that swept through their homes in Malaysia and Java. Similarly, black leopards and black servals demonstrated better resistance, as shown by their survival rates, to viral epidemics in the Aberdare Mountain Range in Kenya.

exquisite sensitivity to touch, pressure, and vibration. His hairless nose pad, also extra thick, can detect temperature differences as little as 1 or 2 degrees. The rest of his body is relatively insensitive to temperature, betraying no sign of pain until skin temperature reaches 126 degrees F. Unlike you, your cat can't cool himself by sweating. Instead, he pants and licks his fur to coat it with saliva, enabling rapid evaporative cooling.

Apocrine sweat glands in each follicle all over your cat's body are not useful for cooling, but they do secrete a milky fluid that's important in sexual attraction. Specialized apocrine glands in the chin, lips, temples, and base of the tail secrete chemicals and pheromones, vital in intercat communication. These are the pheromones that your cat uses to "mark" objects in his environment as his property.

A typical short-haired cat's guard hairs are about two inches long, while a Persian's may reach five inches. The proportion of guard to secondary hairs varies by breed. Devon Rex cats have no guard hairs at all. "Hairless" Sphynx cats actually sport coats of very fine, downy fur.

Coat color and pattern, determined by complex genetics, depend on the presence, location, and proportion of pigments. Every domestic cat is essentially a tabby, though. Your cat's coat, whatever his colors and markings, merely masks the "agouti" pattern of his African wildcat (*Felis silvestris libyca*) ancestors.

Skin and coat problems—dandruff, dry or greasy fur, offensive odors, itching and scratching, failure to self-groom, excessive shedding, hair loss—are often the first symptoms of poor nutrition, parasites, allergies, illness, or psychological problems. See chapter 6 for tips on outsmarting your itchy cat.

Your Cat's Paws and Claws

Your cat's paws and claws form a beautifully integrated system of versatile, efficient tools and weapons.

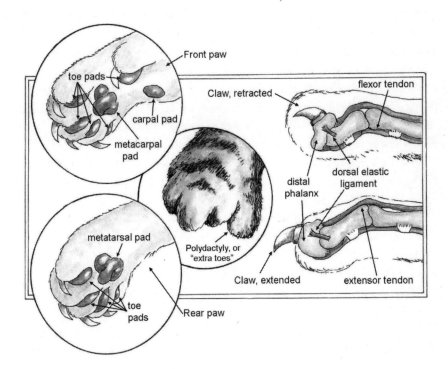

Your cat's marvelous paws and claws.

On most cats, each front paw carries five toes, five toe pads, five claws, a large *metacarpal pad*, and smaller *carpal pad*. The innermost toe, while not opposable, works much like our thumbs in grasping and climbing, while the paw itself can rotate inward to swat or grip prey. Each hind paw carries four each of toes, toe pads, and claws, plus a *metatarsal pad*.

Papa's Polys

At the Hemingway Home and Museum in Key West, Florida, numerous descendants of author Ernest Hemingway's prized polydactyl cats wander the lush tropical gardens and greet visitors. These pampered "conch cats" are named after movie stars, authors, and characters from Hemingway's books. Most of the cats are spayed or neutered, but select polydactyls are allowed to reproduce, to honor Hemingway's fascination with his multi-toed feline companions.

Cats with a genetic mutation called *polydactyly* sport one or more extra toes. They can have up to twenty-eight toes in all. The extra toes, which generally cause no problems, are usually on the front paws but can show up on all four feet. Polydactyl cats look like they're wearing baseball gloves or giant mittens. Polydactyly is very common in some regions, such as New England, and rare in others.

The thick (1/20 inch), tough skin on each foot pad, or *torus*, offers your cat protection on rough surfaces, excellent traction on slippery ones, and a silent stalking tread. Using a clever self-protective strategy, your cat employs his sensitive foot pads to investigate the temperature, texture, size, and shape of unfamiliar objects. First, he'll extend his preferred paw. Then he'll give the object a tentative tap, then a firmer, more confident tap, perhaps followed by a swat or a closer examination with his nose.

As the cat moves about, *Pacinian corpuscles*—sensors within the dermis and fatty layers of the pads—relay information about his orientation and posture to the brain. They're also sensitive to tiny vibrations, working in concert with other senses to help the hunter locate and track potential prey.

Found nowhere else on the body, *eccrine sweat glands* in the foot pads help a hot or frightened cat exude watery sweat—and leave telltale damp footprints. Tucked between the foot pads, *sebaceous glands* secrete a scented oil that keeps the pads moistened and supple and leaves a distinctive scent mark, undetectable to us but clear to other cats, on objects the cat strokes. Once a cat chooses a suitable location to strop his claws, he'll keep returning to it. Read chapter 6 to discover how to outsmart your cat's scratching urges and redirect him to acceptable scratching sites.

Never underestimate your cat's paws. While they may gently caress your cheek, they also conceal a specialized weapon system of awesome power.

Made of *keratin*, the same horny protein that forms the outer layer of the skin, each claw—actually part of the skin rather than the skeleton—is anchored firmly to the terminal toe bone by the *dorsal elastic ligament*. Claws ordinarily lie beneath folds of skin under the toe pads, retracted by ligaments. But when necessary, *deep digital flexor muscles* in the toes rotate the toe bones forward, extending the claws, ready in a heartbeat for defensive or offensive action—or just a good scratching session.

Each claw grows continuously. While most cats keep their claws conditioned and well honed by stropping them regularly, you might want to clip the claws to prevent damage to your furnishings, and your skin. Read chapter 6 to learn more about claws and their care, and how to prevent claw damage.

Foot pads are very sensitive, so many cats dislike having their pads touched or stroked. If you plan to keep your cat's claws trimmed, start in kittenhood to accustom your cat to having his paws, claws, and toes handled gently and respectfully. This will make your care tasks much easier over his lifetime.

Choose a Paw

Is your cat left-pawed or right-pawed? Watch him carefully and see which paw he extends first in tapping and investigating interesting objects. A study of sixty cats by Dr. J. Cole at Britain's Oxford University found that 38 percent of the studied cats were exclusively left-pawed. An additional 20 percent showed a decided preference for using their left paw as their "primary" paw. (Only about 10 percent of humans are left-handed.)

This means that most cats are right-brain dominant, because the right side of the brain controls movement on the left side of the body. In humans, right-brain dominance is associated with a high degree of intuitiveness.

Your Cat's Tongue

Your cat's long, flat, flexible tongue is a marvelously multifunctional instrument.

It pulls food into your cat's mouth, where taste buds embedded in the tongue's surface relay information about taste, temperature, and texture to his brain. Ever wonder why your cat turns up his nose at

Chrysanthemum is very proud of her versatile tongue.

No Candy for Kitties

It's well-known that cats, unlike dogs, don't crave sweet treats. Scientists now know why: Thanks to a genetic mutation, cat tongues are unable to detect sweetness.

We humans possess receptors on our taste buds for five distinct types of flavors: sweet, sour, bitter, salty and "umami." The receptors for "sweet" tastes consist of two different proteins produced under the direction of two genes, and then combined to make a single receptor. When we taste sugar, the receptor tells our brains, "sweet!"

In cats, though, one of the genes that produces the sweetness-detecting proteins is missing some crucial information—so one of the proteins doesn't get made properly. Without both proteins working together, the cat has no way to detect sweetness. But because cats, from domestic felines to lions and tigers, are obligate carnivores, they really don't need to be able to taste sugars. The typical wild feline's diet is rich in protein and fat, and low in carbohydrates and sugars. Therefore, the loss of their "sweet tooth" has never posed a problem for them. As James Richards, DVM, spokesman for the American Association of Feline Practitioners and director of the Cornell Feline Health Center at Cornell University College of Veterinary Medicine, once put it, "A sweet receptor in cats would be like me having a taste receptor for wood."

food right out of the refrigerator? He prefers his food closer to 86 degrees F, his tongue's normal temperature.

It's also a very sophisticated water-dipper. Unlike dogs, who slurp noisily, cats sip quietly and neatly, never getting their chins or whiskers wet. The clever cat simply touches the surface of the liquid with the smooth tip of his tongue, rapidly pulls up a column of liquid and then—when the column is as high as it can be before falling back down—closes his jaws and captures the liquid. (See "Learning How Cats Drink," page 34.)

It's a cleaning appliance, smoothing ruffled fur, drying wet fur, and distributing saliva (with its built-in cleansing and deodorizing agents) to every part of his coat. Powerful, rearward-facing, hook-like barbs along the

Watch That String

Linear materials like yarn and string are dangerous cat toys. Once your cat has yarn or string in his mouth, the barbs on his tongue make it nearly impossible for him to spit it out. If swallowed, yarn, string, and similar materials can cause serious internal injuries.

center of the tongue, called *filiform papillae*, help your cat remove dirt, dust, and other foreign matter from his coat.

It's a rasp. The filiform papillae come in handy for stripping meat from bones.

It's an air conditioner, too. An overheated cat curls his tongue into a tube for efficient panting, and distributes saliva over his coat, enabling rapid evaporative cooling.

Your Cat's Nose and Vomeronasal Organ

Ever see your cat with his head cocked at a slight angle, eyes half closed, mouth narrowly open, lips curled back in what looks like a grimace or snarl, maybe even drooling?

He's "smell-tasting," using a specialized organ between the roof of his mouth and his nasal cavity called the *vomeronasal* or *Jacobson's organ* to sample a par-

ticularly interesting odor. His weird expression is called the *flehmen reaction* or *gape*. He's flicking the tip of his tongue against his upper palate and the roof of his mouth behind his teeth, to better deliver the odor molecules to his vomeronasal organ. Contrary to appearances, a flehmening cat is not grimacing in disgust, but analyzing and savoring a chemical message that carries extraordinary pleasure or importance. By flehmening the messages in

Silver "flehmening."

a urine mark, for example, your cat can discover the marker's gender, sexual availability, status, territoriality, and agenda.

Cats also flehmen when "sampling" the odors of feces, gland secretions, certain plants, and a variety of nonbiological scents, including many common cleaning products.

Your cat's olfactory system has about 200 million odor-sensitive cells, while you have only about five million. He uses scent as his primary means of recognizing

Learning How Cats Drink

Cats and scientists, with their endless curiosity about the world, have long enjoyed fruitful partnerships. In the seventeenth century, the great mathematician Sir Isaac Newton, inspired by his cat, invented the cat-flap. In the twenty-first century, another curious scientist also inspired by his cat discovered how cats drink.

Dr. Roman Stocker, a biomechanical engineer at the Massachusetts Institute of Technology, usually studies how microorganisms like plankton move about in their tiny, watery worlds. But one day, as he was watching his cat, Cutta Cutta, lap water, he realized he was observing a very interesting biomechanical phenomenon. Had anyone actually studied the biomechanics of cat-lapping? As it turned out, no one had.

Unlike us, cats can't seal their cheeks to take in liquids via suction. (Neither can dogs.) Even though cats sip neatly, quietly, and elegantly in comparison to noisily slurping dogs, it has long been assumed that their drinking techniques were probably similar; both, it was thought, curved their tongues into ladles and scooped water into their mouths. But cats, especially, lap so rapidly that no one could tell for sure.

One of the specialties of Stocker's MIT lab is micro video photography, so he had the right tools at hand. He and his colleagues (Pedro M. Reis of MIT, Sunghwan Jung of the Virginia Polytechnic Institute, and Jeffrey M. Aristoff of Princeton University) filmed Cutta Cutta lapping. The researchers first thought that the rearward-facing barbs on the cat's tongue (the *filiform papillae*) might have something to do with the mechanics of drinking. But it turned out that only the smooth tip of the feline tongue is involved.

Here's how it works: The cat curves the upper side of his tongue downwards into a "J" shape, so that just the smooth tip delicately touches the water's surface. The surface tension of the water makes the liquid stick to the tongue. The cat then pulls up, at high speed, a column of water. At the exact instant when the column is tallest, and the exact instant before gravity would pull the column back down, the cat closes his mouth and captures the liquid. Neat! In scientific terms, the cat's lapping method depends on his instinctive ability to calculate the point at which gravitational force would overcome inertia and cause the water to fall.

The scientists discovered that cats do indeed lap quickly: four times per second, much too fast to see with the unaided eye. The tip of the cat's tongue moves at a speed of about three feet per second. They also calculated how much liquid a cat like Cutta Cutta takes in with each lap: about 0.1 milliliter per lap. At four laps per second, that's nearly five teaspoons per minute.

So that they could make additional measurements and observations without further inconveniencing Cutta Cutta, Stocker and his colleagues built a simple

Cutta Cutta, Inspirational Cat of Science
Micaela Pilotto, Pedro Reis, and Roman Stocker

robotic tongue: a piston capped with a glass disk about the size of the tongue tip. They wanted to see exactly how fast a cat would have to lap to maximize his water intake. The engineers experimented with different lapping speeds and positions, and discovered that cats lap at precisely the most efficient speed. Cats instinctively know just when to close their mouths to get the most water.

Next, the researchers devised a mathematical model of cat-lapping. Their model stated that "the lapping frequency should be the weight of the cat species, raised to the power of minus one-sixth and multiplied by 4.6." (I will spare you, and myself, the equations.) Simply, the model predicted that the larger the mass of the cat, the slower the lapping would proceed.

Now they needed to observe a wider variety of felines drinking. At the Franklin Park Zoo in Boston, Massachusetts, and the Stone Zoo in Stoneham, Massachusetts, the scientists took high-speed videos of lions, ocelots, leopards, and jaguars. Sure enough, all these cats lapped at just the speeds predicted by their model.

For further data, the researchers turned to YouTube and other online video sites, with their millions of cat videos. They found that the cat-drinking speeds in these films also adhered to their mathematical model's predictions.

This very elegant and interesting piece of research was published in the online journal *Science* (November 11, 2010) and was reported in the journal *Scientific American* and on the front page of the *New York Times,* ensuring scientific immortality for Cutta Cutta, a gray-and-white rescue from a Boston-area shelter. And it proves once again that when it comes to cats, there's always something more to learn.

individuals, objects, activities, opportunities, and threats in his environment. With his sensitive sniffer, your cat keeps track of news, warnings, and local traffic, and gathers information to help him pursue prey, entertainment, and romance. His nose warns him about dangers (spoiled food, a lurking dog); announces opportunities (a receptive female, nearby prey); and reveals the activities and agendas of other cats.

Read chapter 5 to learn how to live harmoniously with a creature endowed with such an exquisite sensitivity to odors, and how to outsmart the problem behaviors that can result.

Your Cat's Vocal Cords

Your cat "speaks" by vibrating his vocal cords, elastic bands of muscle that lie within his larynx, a circular organ made of cartilage and muscle in the throat at the opening of the *trachea*, or windpipe. His voice box, or *larynx,* is connected directly to his skull by the *hyoid bone.* The nature of this structure prevents your cat from producing the ground-shaking roars that big cats like lions can make, but allows him to purr continuously while inhaling and exhaling. At the top of the larynx is a leaf-like flap called the *epiglottis.* It automatically closes when your cat swallows, keeping food from going down his windpipe.

Read chapter 10 to learn more about your cat's vocal repertoire, from growls to purrs to the "silent meow," and how to outsmart a too-talkative, or too-quiet, cat.

Your Cat's Ears

Your cat's ears collect, focus, and deliver sound vibrations to his brain, providing him with his sense of hearing, while the *vestibular apparatus* of his inner ear helps him maintain his balance. The fluid-filled structures of his inner ear—the looping *semicircular canals*, the *utricle*, and the *saccule*—are essential to his self-righting ability. Every movement causes the fluids in his vestibular structures to ripple the millions of microscopic hairs that line them. These tiny ripples, along with the movements of fine particles that float within the utricle and saccule, flash to your cat's brain precise, up-to-the-microsecond information about his actions, position, and orientation in space.

Your cat can easily perceive sounds at much higher frequencies than humans or even dogs—up to 50 to 65 kilohertz. (Our limit is about 18 to 20 kHz.) The

Is Your Cat Psychic?

Extraordinarily sharp senses, mysterious nocturnal habits, and solitary, stealthy natures have long given cats a reputation for having a "sixth sense," able to perceive unseen or unearthly and paranormal forces. Many cats have made documented, incredible journeys back to cherished homes or people. During the Middle Ages, this reputation for paranormal connections, psychic powers, and "psi-trailing" cost millions of cats their lives, because such powers were associated with the devil. Today, though, it's given rise to a whole new specialty: "pet psychics" and "animal communicators." For a fee, they'll contact your cat's mind and find out what he's thinking and what's bothering him.

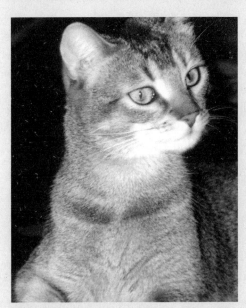

Is Tabigail psychic? Or just very, very observant?

Pet psychics' services cost anywhere from $35 for a fifteen-minute phone consultation to several hundred dollars per hour. For the price of a few sessions, you could get a deluxe combination scratching post and climbing tree with all the trimmings.

I remain skeptical—although when it comes to cats, I rule nothing out. If my cat has something important to say, why would he disclose it to a stranger over the phone? If you believe that you can access your cat's mind directly, try it. Develop your own intuitive powers. Talk to your cat and *listen* to him.

Meanwhile—I'll go with the climbing tree.

shape, mobility, and focusing facility of his external ear structures allow him to detect extremely faint sounds—like the tiny, high-pitched squeaks of mice. All this also helps explain the heightened attention your cat pays high-pitched human voices and kittenish mews, and his aversion to loud or harsh noises.

Your cat's inner ear gives him an exquisite sensitivity to static electricity, sounds, and vibrations imperceptible to humans, and to tiny fluctuations in atmospheric pressure and in the earth's magnetic field. This sensitivity has given the cat, throughout the ages, a reputation as an uncannily accurate predictor of storms and earthquakes.

Before humans understood the science behind many atmospheric and geological phenomena, the cat's "psychic" weather- and disaster-forecasting abilities were attributed to magic or to a mysterious sixth sense, and gave rise to much folklore and numerous proverbs. But it's probably just the tickling of his finely tuned inner ear structures in response to such stimuli that causes cats to scratch or wash their ears more vigorously before a thunderstorm, or to display unusual agitation as the earth subtly vibrates.

The nineteen muscles in each of your cat's *pinnae*—the visible, cone-shaped, outer ear flaps made of delicate, skin-sleeved cartilage—allow him to swivel his ears independently up to 180 degrees around and prick them directly toward a sound source much faster than any dog could. The separation between the pinnae and their ability to swivel independently lets your cat pinpoint the location of a sound with singular precision. He can distinguish sounds 5 degrees apart—only about three inches apart at a distance of three feet.

Older cats, like elderly humans, can lose much or all of their hearing. For a solitary feline predator in the wild, seriously diminished hearing can mean the difference between survival and starvation. But in safe, familiar surroundings, even a completely deaf cat can enjoy a long, happy life.

Some cats are born deaf. White cats with blue eyes are much more likely to be affected, due to a congenital absence or degeneration of the inner ear's organ of Corti. A white cat with only one blue eye will likely be deaf only in one ear, usually on the blue-eyed side.

Ear infections, *otitis* (inflammation of the ear), and parasites such as ear mites can cause hearing loss as well as tremendous misery. If your cat suddenly starts digging and clawing furiously at his ears, shaking and tilting his head, squinting in distress, or walking or running around in circles, or if you see dark, gummy or other unusual material in his ears, call your veterinarian immediately.

Your Cat's Whiskers

Each of your cat's whiskers is an exquisitely sensitive receptor, thick and stiffened at the base, tapering to dainty fineness at the tip, yet firm enough to brush against obstacles without collapsing. Growing from a distinctive follicle strengthened by a fibrous capsule, each whisker is twice as thick as your cat's guard hairs and extends three times deeper into his skin. The follicles lie in a bed rich with blood vessels and nerve endings. Connected to the follicles are *arrector muscles* that allow your cat to instantly pivot his whiskers. Each whisker is *innervated*—connected to nerves that transmit impressions directly to the brain—where visual and tactile centers occupy adjacent areas.

Your cat uses his whiskers as both a rough gauge of the size of a passage, and an early warning system of what lies beyond. Whiskers transmit information carried by tiny air currents directly to his brain, enabling him to evade potential dangers or pursue interesting opportunities. By detecting subtle changes in atmospheric pressure and sensing air currents around obstacles, whiskers enable swift, confident navigation, even in darkness. Along with his other senses, especially vision, whiskers keep your cat informed about what's nearby and where he is in relation to it.

Whiskers also help your cat avoid eye damage. Potential dangers graze his whiskers first, triggering a protective blink. Even in dim light, when he can't focus on close objects, whisker data let him proceed with speed and confidence. Whiskers verify and reinforce information gleaned through his other senses, and provide critical data obtainable in no other way.

Your cat's most prominent whiskers are his *mystacial vibrissae*, arrayed across his muzzle (cheeks) in four horizontal rows, with approximately twelve whiskers on each side. The strongest, thickest whiskers are in the two middle rows. He can move the top two rows independently of the bottom two.

The tufted *genal vibrissae* extend from high on the outer edges of your cat's temples. His *superciliary vibrissae* arch gracefully above his eyes, while his *carpal vibrissae* sprout from the backs of his forelegs. These "paw whiskers" are a common feature of carnivores who use their forelegs to grasp prey. Many cats also have short, bristly, whisker-like hairs on their chins that, while not true whiskers, possess enhanced sensitivity.

When he captures prey, your cat needs to know whether it's still alive and exactly how it's positioned, so he can apply an accurate killing bite. The carpal whiskers, in concert with the mystacials, fanned forward in an information-gathering net, tell him.

Whisker Charms and Whisker Weapons

Many animals have whiskers, or vibrissae, but it's the singular charisma of feline whiskers that's captured the imagination and fired the curiosity of shamans, scientists, and owners trying to outsmart their cats. Feline whiskers have long been steeped in mystery and ascribed formidable supernatural powers.

To honor the jaguar, whom they revere as their ancestor, the Mayoruna Indians of the Brazilian rain forest traditionally implanted slivers of palm straw into their noses in patterns vividly reminiscent of feline whiskers. This traditional tribute died out as the Mayoruna forged contacts with outsiders. But many young men and women still have long parallel lines—jaguar whiskers—ritually tattooed across their faces.

Tradition in India, Malaya, and China held that singeing or plucking a dead tiger's whiskers deprived the cat's spirit of its navigational ability, rendering it less dangerous to the living. Removing the whiskers also prevented the soul of a man-eating tiger from passing to another killer tiger.

In Malaya, a single tiger whisker ground up into food was said to have the power to kill a man. In Indochina, a whisker placed in a shoot of bamboo condemned your chosen enemy to an agonizing death. In Indonesia, tiger whiskers—burned, rubbed into a powder, and mixed with food—were a surefire remedy for impotence.

Curators at museums that feature displays of animals preserved through taxidermy must often replace the whiskers of their tigers. Some museum visitors—with a score to settle or in need of a miracle cure, and unaware that the tiger's original whiskers are long gone—filch the realistic plastic replacements.

Most cats dislike having their whiskers handled, especially against the direction of growth. Never clip, trim, or remove these important sensors; they take months to grow back. Seventeenth-century author Edward Topsell warned, ". . . if the long hairs growing about her mouth be cut away, she losseth her corage." Your cat sheds whiskers periodically; look near spots he cheek-rubs. Save them as magic charms.

Your Cat's Eyes

In relation to body size, cats have the largest eyes of any mammal. Like most predators, your cat has forward-facing eyes and binocular vision—a sweeping, 285-degree field of view, including 130 degrees of overlapping vision and 75 to 80 degrees of peripheral vision on each side. The fusion in your cat's brain of the slightly different images from each eye enables him to see in three dimensions and to accurately judge distance, depth, and size.

"Stalk-and-pounce" predators rely on a symphony of sensory input to locate and kill prey. Your cat's vision is exquisitely adapted to the needs of a *crepuscular* (dawn and dusk) hunter. Old tales that cats can see in pitch darkness are false, but cats are able to take advantage of 50 percent more available light, and have six times better night vision, than humans. Your cat's eyes are perfectly adapted for low-light hunting:

- Eyes large in proportion to head size

- Pupils large and nearly spherical

- Cornea and lens set close to the eye's center

- Lens set close to the retina to concentrate light

- 25:1 ratio of rods (sensitive to low-intensity light) to cones (sensitive to bright light), in contrast to our 4:1 ratio

The *tapetum lucidum*, a mirror-like, dense layer (up to fifteen cells thick) of highly reflective zinc and specialized proteins lining the back of the eye, harvests and re-reflects light back to restimulate the retina's rods. This gives your cat's eyes a second chance at all available light.

Many rods double as motion sensors, so your cat can detect movement much better than you can. Motion *across* your cat's field of vision registers much quicker than motion *toward* him. Movement of the right kind *away* from your cat triggers predatory pursuit—something to remember during interactive play.

In dim light, with his pupils widened to balls, your cat's vision is somewhat fuzzy, and he can't distinguish fine detail. Even in daylight, he can't focus well up close; his vision is sharpest from about 6½ to 20 feet. But as hunters of dusk and dawn when colors are subdued, cats have little need to distinguish colors. They can distinguish several shades of gray, blue, green, and yellow. Red likely appears gray.

Your cat's "third eyelid," also called the *haw* or *nictitating membrane*, is an opaque fold of skin between his lower eyelid and the inner corner of his eye. It serves as a shield against glare and also helps moisten the eye and clear dust from the cornea. If you see more than a sliver of the third eyelid in your awake cat, schedule a visit to your veterinarian. It could indicate illness.

Rather mysteriously, cats blink only infrequently—as little as once every few minutes. Writer Anitra Frazier fancies that cats communicate in blinks. Gaze into your cat's eyes, she suggests, and blink three times, slowly, while thinking, *I . . . Love . . . You.*

Decoding Feline Body Language

POSTURE AND GAIT

Your cat's posture and gait reveal a lot about what he's thinking and feeling. A "Halloween cat"—back arched, head lowered, tail down, fur fluffed—is scared, but trying to look larger and fiercer. A cat facing sideways, up on tiptoes, is hoping his large apparent size will cause his opponent to slink away quietly.

A cat who crouches or slinks past another cat, tail tucked down, ears lowered, and eyes averted, wants that other cat to ignore him. A cat who's confused, uncertain, puzzled, or embarrassed might look away and nonchalantly lick a paw, or carefully groom his already perfect tail. He's "displacing" his nervousness.

Is your cat galloping about, leaping onto bookshelves, and merrily scattering papers? Prancing, tail waving high, head proudly erect, ears pricked, eyes shining? He's a happy cat—confident, pleased with the world and himself. Is he on your lap, backing up rear-end-first to your face, tail waving high? He thinks you're his mom, and he's presenting his rump for inspection and cleaning, like small kittens do. This is an indication of his love and his trust in you.

Kitty Rollover

In one of the most misinterpreted cat postures, your cat rolls over and exposes his tummy. Don't assume he wants his tummy petted, though. Unlike dogs, who roll over to show submission to a dominant animal, a cat on his back is poised for defense or offense, with his full array of weapons pointed up and ready for action.

Many cats use this charming posture simply to display friendship, love, and trust—but really don't like their tummies touched. It's not that he doesn't love you, but his tummy is *very* sensitive, and petting or even touching it can over-stimulate him. See chapter 9 for tips on outsmarting an overstimulated cat.

Your cat has special scent glands on his forehead, lips, and cheeks, and at the base of his tail. These glands secrete *pheromones*—your cat's personal "signature

Is Dandelion begging for a tummy rub? Or just showing off her tabby "weskit buttons"?

scent." When he slinks around your ankles, bunts your head or face, nuzzles into your hand, or pushes his cheek against your leg, he's scent-marking you to claim you as his property, and as an important feature of his territory. That's right—you're part of his territory. Consider this a compliment. It means you're an honorary cat, accepted as a member of your cat's preferred social circle. You can't detect the scent mark, but other cats can.

TAIL TALK

> "His tail is the mirror of his mind."
> —*Walter Chandoha, author and photographer*

Many critters communicate with their tails, but the cat turns tail talk into art. His tail can coax, flatter, tease, beguile, lead, demand, amuse, warn, manipulate, and charm. Writer Clair Necker observes that cats flick their tails in situations where humans would thumb their noses.

Your cat speaks his own tail dialect, but all cats share a common vocabulary. A tail held high is a greeting, indicating alertness, attention, and equanimity. A relaxed cat strolls with his gently waving tail raised to about a 40-degree angle. As he pauses to investigate a tantalizing scent, his tail twitches inquisitively. He spies a mouse, and the chase is on! As he gallops, his tail flattens and streams behind, ready to perform its spontaneous balancing magic.

With a sudden and dramatic fluffing of his tail, a fearful or threatened cat can make himself look larger just long enough to escape an unpleasant confrontation. If immediate escape seems impossible, he holds his now bristling tail high to signal his unwillingness to give in, or arches it into an inverted "U" to assure his opponent that he's ready to fight if necessary.

A slow, deliberate lashing of the tail means a seriously annoyed cat. The degree and vigor of the lashing indicate how mad he is. Unlike the wagging tail of a friendly dog, a thrashing feline tail is a clear warning: *Back off, now!* Two cats who approach each other with tails thrashing are spoiling for battle, or at least intimidation.

In his low-to-the-ground hunting slink, your cat lashes only the very tip of his tail. This helps him contain his growing excitement, defuse his impatience, and fine-tune the positions of his muscles and spine for the final pounce.

Tail talk serves as feline social shorthand. A cat approaching a nonthreatening feline acquaintance will raise his tail politely in greeting. In a multicat household, the top cat may feel obliged, or entitled, to raise his tail higher than the subordinate cats he greets. (Dominant lions, tigers, and leopards show a curved "scorpion tail.") A cat who holds his tail closely tucked in, especially when his paws are tucked in, too, likely wants to be left alone. A twitchy, nervous tail is the sign of an edgy, uncertain cat.

WHISKER WHISPERS

> "... few animals display their mood via facial expression as distinctly as cats. The cat's face invariably reveals what it is up to and what kind of behavior will occur in the next moment."
> —*Konrad Lorenz, ethologist*

Your cat can dodge and swivel his whiskers with lightning speed to express emotions, sense subtle shifts in air currents, and navigate swiftly through darkness. He can instantly pivot his whiskers, swing them to alertness, fan them forward, or sweep them back. When he tilts his head to sniff an intriguing odor, his mystacial whiskers lie retracted gently against the sides of his face. While resting or relaxing, he extends them out sideways.

If he's anxious or alert, he'll arch his whiskers gently forward and upward. Strolling or exploring, he fans them farther forward, perhaps twitching or circling them if it's dark or he's on unfamiliar turf. When annoyed or threatened, he flattens his whiskers tightly back against his cheeks as his ears pull forward and his eyes narrow. In the excitement of pursuit and capture of prey, his mystacial whiskers fan fully forward, forming a kind of net in front of his mouth. Toss him a treat and see how quickly his whiskers fan forward to "net" it.

Sometimes an exquisitely delicate brush of whisker tips (a "cat kiss") replaces the familiar facial bunt-and-rub between two friendly cats. An old folk belief holds that mother cats chew off their kittens' whiskers to keep the babies from straying, or to show affection for their favorites.

EAR SIGNALS

Each of your cat's *pinnae* (cone-shaped external ears) can independently pivot, swivel, and rotate into endless expressive configurations. Erect and rotated forward in alertness, flattened in defense, rotated sideways in aggression, twitching

Bunny exercises all thirty-plus muscles in her ears to show her feelings.

in frustration, or just relaxed, the ever-shifting positions of your cat's ears reliably mirror his mood, intent, and feelings. A puzzled, confused, or uncertain cat often holds his ears in two different positions—one flat out sideways, like an airplane wing; the other swiveled into a quirky point.

EYE CLUES

"It is in their eye that their magic resides."

—*Arthur Symons*

Your cat's pupils are fully dilated when he's apprehensive or terrified. When he's feeling aggressive or predatory, they narrow to vertical slits. They narrow in bright light to prevent dazzle and eye damage. The eyes of a playful, happy cat sparkle with mirth. Half-closed eyes mean a sleepy or very relaxed cat.

Silver prides himself on his expressive eyes.

Quickly Decoding Your Cat . . .

Happy cat greeting you

- "Banner tail" held high

- Quick tail flicking

- "Kitty dancing"—a cute little hop, skip, jump

- Bunting your head or face

- High-pitched, kittenish squeak or mew

- Eyes wide and friendly

- Slinking and winding around your ankles

- Rafter-rattling purr

Happy cat greeting a feline buddy

- Cats hold tails upright, swaying gently

- "Whisker kiss"—brief touching of whiskers, or nose tap

- Top-of-the-head "love lick"

- Full-body pass, side by side

- Same movement, but cheek-to-cheek

- Sniff or brief lick just above tail

Cat meeting unfamiliar cat

- Cats check each other out from several feet away

- "Airplane ears" (flattened and aimed sideways)

- "Cat humming"—elaborately disinterested paw licking or grooming

Scaredy-cat

- Pupils huge and round as balls

- Paws and tail tightly tucked in

- *Piloerection*—fur fluffed out, especially on tail and along spine

- Crouching posture

- Airplane ears, *or* ears flared a bit backward

- Swept-back whiskers; uncertain expression

- Very low growl, almost a rumble; *or* whine or hiss; may escalate

Aggressive cat—on offense

- Tail vigorously thumping or lashing rapidly

- Exaggerated, spiky piloerection

- Crouched low, rump swinging side to side—ready to spring

- Ears and whiskers plastered flat backward against head

- Snarl: showing teeth, ready for action

- Spit: an extreme hiss

- Extending claws to show their sharpness and readiness

- Huge, round pupils

- Sound effects in terrifying combinations: scream, whine, growl, rumble, hiss

Cat on the defense—looking to escape, but willing to fight

- Extreme piloerection—*Watch out: I'm a huge cat!*

- Tippy-toe, sideways stance, facing opponent but looking for escape opportunities

- Huge, round pupils, eyes darting around, looking for an exit

- Flattened whiskers and ears, or airplane ears

- Sound effects designed to stave off attack—wailing *Meeeooooow!* or hiss

Confused, anxious, or uncertain cat

- Cat humming—trying to look unconcerned by exaggerated grooming

- Ears twitching asymmetrically, or held flattened and sideways

- Nervous twitching of tongue and tail

- Soft, kittenish mews or squeaks

Outsmarting Feline Illness

Your cat counts on you to ensure that he remains in top physical condition and health. The two keys to outsmarting feline illness are *preventive care* and *everyday awareness.* Your number-one partner in keeping your cat healthy is your veterinarian, who has the specialized knowledge and expertise to diagnose and treat a wide variety of feline ailments. But nobody knows your cat as well as you do, and when it comes to feline illnesses, preventive care is always less expensive—and easier on you and your cat—than treatment. By integrating these eight simple prevention and awareness techniques into your daily life, you can help your veterinarian outsmart almost every feline health problem that might arise:

1. Provide your cat with regular veterinary care, including annual or semi-annual "well-kitty" checkups.

2. Make sure your cat gets all appropriate vaccinations, tailored to his individual circumstances.

3. Take good care of your cat's teeth, because many health problems begin in the mouth.

4. Maintain your cat at a healthy weight.

5. Give your cat quick at-home checkups at least once a week.

6. Keep a watchful eye on your cat every day so that you'll notice small problems early, before they turn into big ones.

7. Be alert for the telltale signs of feline illness.

8. Look in the litter box for early clues to illness.

REGULAR VETERINARY CARE

Regular veterinary care is a sound investment in your cat's health and happiness. No matter how well you care for your cat at home and how observant you are, no at-home care can substitute for regular veterinary care. All cats need to visit their veterinarian at least once a year for a complete checkup. Senior cats (age eight and up) and cats with chronic medical problems should see their veterinarian at least twice a year. Well-kitty checkups save lives.

Cats Can Get "Alzheimer's," Too

For almost twenty-one years, I lived with an extraordinary black cat named Dominique. "Nika" was unusually quick-witted, intelligent, curious, and challenging—even for a cat. She was my "proof of concept" cat: All her life she did things—sometimes quite maddening things—just to prove they could be done and that she could do them.

As she got older, she slowed down a bit and stopped challenging me at every turn (and believe me, that was a relief!). But I wasn't really prepared when she started awakening in the middle of the night, howling in apparent distress and confusion, and wandering around the house, calling out, looking puzzled and lost. I took her to the veterinarian for a complete checkup. The diagnosis: feline cognitive disorder, or "Kitty Alzheimer's."

Like humans, cats can't be definitively diagnosed with Alzheimer's disease while they're still alive; it's only after postmortem examinations are carried out that the diagnosis can be confirmed. Necropsies show that the brains of cats who had shown symptoms of age-related cognitive disorder contain the same kind of thick, gritty, tangled fibers, called beta amyloid plaques, that also characterize the brains of deceased human Alzheimer's patients. Studies have shown that 28 percent of cats between ages eleven and fifteen and 50 percent of cats over age fifteen show some signs of cognitive dysfunction.

It makes sense, really. As we've been able to offer our cats safer lifestyles, better nutrition, and improved preventive and other veterinary care, they've responded by living longer. But these longer life spans also put them at risk for the same kinds of age-related illnesses humans suffer from, including arthritis, various cancers, diabetes, and cognitive decline. At the time I started noticing cognitive behavioral changes in Nika, she was about sixteen—the equivalent of eighty years old in human terms.

Fortunately, Nika was a disciplined and reliable litter box user right up to the end of her life. But many elderly cats start missing the box—maybe forgetting where it is, perhaps not getting there in time every time. Although this can be frustrating, remember that punishment never works—and is downright cruel in this situation.

Other signs that your aging cat may be experiencing cognitive decline include extreme or profound lethargy; loss of interest in activities she once enjoyed, such as interactive play; radical changes in her normal patterns of sleep and wakefulness; failure to show up at mealtimes; staring into space or staring at walls for long periods of time; greatly increased need for sleep; uncharacteristic hiding; apparent failure to recognize familiar people or animals, possibly leading to conflicts and fights; uncharacteristic aggression; agitation that seems to have no cause; howling, especially upon awakening; and apparent disorientation. Any of these signs, especially in combination, call for a veterinarian visit.

Your veterinarian may suggest changes in diet to help control age-related illnesses (such as diabetes or kidney failure) that many elderly cats suffer from. There are also some new prescription medications that have been shown to help slow down (but not stop or reverse) your cat's cognitive decline. One such drug commonly and successfully used in elderly cats is Anipryl (selegeline). Other medications and supplements that might help include antioxidants, sertraline, lorazepam, melatonin, L-theanine, and pheromones. Each cat is an individual, though, and it may take awhile to find a combination that helps.

Some cats age more gracefully than others, and many reach ripe old ages without ever showing signs of cognitive decline. Sometimes, though, major changes are called for. Because feline cognitive disorder can affect memory, learning, alertness, and awareness of surroundings as well as behavior, an outdoor or indoor-outdoor lifestyle is extremely risky. Cats who have shown signs of age-related cognitive problems should be kept indoors exclusively. Even cats who have always enjoyed pampered, indoor-only lifestyles need and deserve extra attention paid to their environment to ensure that their quality of life remains as high as possible.

If your aging cat is showing signs of cognitive decline, your job is to offer gentle, discreet assistance so that she can maintain a high quality of life. Here are some steps you can take to make your cat's senior years safer and happier:

- Move and alter litter boxes to make them easier to find and access.

- Add extra litter boxes so that there is at least one on each level of your home. Don't make an elderly cat go down the cellar stairs to get to her box!

- Provide plenty of warmth, as elderly cats are often too cold. Use cat-safe heating devices in your cat's bed.

- Don't stop daily interactive playtime, but alter it so that it better suits your cat's abilities. She may not want to jump and run, so offer alternative, gentler play sessions.

- Offer gentle grooming assistance. Although Nika was always a particular and fastidious cat, she eventually reached the point where she was too weak to groom herself effectively or manicure her own claws. I took over, brushing her daily, clipping her claws as needed, and bathing her with a soft, warm cloth.

- Add a cat fountain. The soft bubbling sounds of the water can remind your cat to drink more frequently—especially important for the health of aging kidneys.

- Don't forget mental stimulation! Give your elderly cat plenty to do, watch, and think about (such as treat and food puzzles, or a bird feeder outside a favorite window).

- Try some clicker training. Most cats love it, and learning new behaviors can help keep their minds sharp.

- Consider offering your cat a space of her own: her own room, fitted out with everything she needs. This can be especially kind if you have a busy household that includes other cats and animals, kids, and lots of activity.

Remember, your elderly cat will appreciate, and thrive best in, a stable, predictable, low-stress environment. Regular mental stimulation and environmental enrichment strategies such as daily playtimes, clicker training, and gentle walks around the house (or neighborhood) in harness and leash can help keep kitty minds engaged and sharp. And never forget what my Nika taught me: Old friends are best.

Special Care for Senior Cats

When your cat turns about eight years old, his veterinarian will likely take a set of blood tests called a "baseline blood panel." The results from these blood tests will form a record of what's normal for your cat when he's healthy. These records can be compared against later tests, helping your veterinarian detect early signs of disease. As your cat ages, he becomes more prone to several diseases, including hyperthyroidism, diabetes mellitus, heart disease, various cancers, arthritis, chronic kidney disease, and feline cognitive disorder ("Kitty Alzheimer's"—see box above). Rapid advances in veterinary medicine mean that all of these conditions can be successfully managed (if not cured) with medication, diet, and lifestyle changes, offering your cat his best chance ever at a long, comfortable, and happy life.

VACCINES ARE VITAL

One of the smartest things you can do to outsmart feline illness is to ensure that your cat receives the vaccinations he needs. Modern vaccines are extraordinarily safe and effective, and appropriate vaccinations are vital to your cat's health and well-being. Which vaccinations your cat needs, how they are administered (intranasally or via injection), and how often they should be given depend on a number of factors, including your cat's age, health, medical and behavioral history, and lifestyle.

When you take your cat or kitten for his well-kitty checkup, your veterinarian will first ask lots of questions about your cat's individual life, health, habits, and circumstances. Then you can decide together which vaccinations your cat really needs. Some are essential for optimal health, some are required by law, and others are optional, depending on your cat's individual needs and situation. The American Association of Feline Practitioners (AAFP) has issued guidelines to help veterinarians and cat owners determine which vaccinations each cat should get, and how often. (See box on next page.)

OUTSMARTING DENTAL DISEASE

According to the American Veterinary Medical Association, dental disease is the most frequently diagnosed health problem for cats, showing up in about 70 percent of cats by the time they're only three years old. Regular attention to your cat's dental health is one of the most important things you can do to help ensure him a long, healthy, happy life.

Bad teeth and swollen gums *hurt.* Sometimes these signature signs of *periodontal disease* hurt so much that the cat can't eat, drink, or even swallow without pain. Cats who don't get relief sometimes stop eating altogether, or cut back so much that their overall health is seriously affected. But pain is just the beginning. Bacteria from infections in the mouth can enter the cat's bloodstream, triggering potentially severe infections in major organs such as the liver, kidneys, and heart, and even kicking off long-term problems such as diabetes and chronic kidney disease.

Get your cat gradually and gently accustomed to having his mouth touched and handled. As you're able to, take a close look at the condition of your cat's teeth and mouth. Look for any red or swollen patches on his gums; swelling in his throat, cheeks, face, or mouth, or on his tongue; or broken or cracked teeth. All call for veterinary attention. Be alert for loose or missing teeth. It's never normal for a cat to lose adult teeth. Painful *odontoclastic resorptive lesions*, also called *neck* or *cervical line lesions*, can cause teeth to break off at the gum line.

Check for large buildups of plaque or tartar on the teeth. Saliva plus food forms *plaque*—a sticky substance that clings to teeth and gums. Plaque hardens into *tartar*, a hard, brownish yellow deposit that collects around the baselines of the teeth, especially rear teeth, when they're not cleaned regularly. Plaque and tartar combine

Vaccine Guidelines from the American Association of Feline Practitioners (AAFP)

Core Vaccines

The following vaccines are recommended for every cat:

Panleukopenia Virus (FPV), Feline Herpesvirus-1, and Feline Calicivirus (FHV-1/FCV)

- Initial series for kittens: First dose as early as six weeks, then every three to four weeks until sixteen weeks.

- Initial series for adolescent (older than sixteen weeks) and adult cats: Two doses, three to four weeks apart.

- Boosters: One dose one year after the last initial series dose. Then no more frequently than one dose every three years.

Rabies

- Initial for kittens: First dose at eight or twelve weeks, depending on product label. Revaccinate one year later.

- Initial for adolescent and adult cats: Two doses, twelve months apart.

- Boosters: Annual booster *or* every three years, depending on vaccine, or as required by state or local laws.

---

with bacteria to cause infection, enamel breakdown, loosening, and pain. Pockets of infection and pus in the mouth can make eating highly unpleasant. A depressed appetite can cause your cat's overall health to suffer and quickly lead to serious complications.

Sniff anything? Minor "tuna breath" is normal, but persistent or strong mouth odor is cause for concern. Bad breath can indicate gingivitis or periodontal disease, cancer, or a mouth tumor. Ammonia odor can indicate kidney failure.

If your cat's teeth aren't sparkling white, if you see a lot of tartar buildup, or if your cat has persistent foul breath, it's time to start regular dental care. Neglected,

Non-core Vaccines

Give these vaccines *only* when circumstances indicate.

Feline Leukemia Virus (FeIV)

Recommended for all kittens, and for all cats at risk for exposure. *Only FeIV-negative cats should be vaccinated; test first.*

Feline Immunodeficiency Virus (FIV)

Should be restricted only to cats at high risk of infection. Ask your veterinarian if the FIV vaccine is right for your cat.

Chlamydophilia Felis

Should be reserved as part of a control regime for cats in multicat environments where infections associated with clinical disease have been confirmed.

Bordetella Brochiseptica

Should be considered *only* where cats are likely to be at specific risk of infection. Ask your veterinarian if this vaccine is right for your cat.

Not Generally Recommended

Feline Infectious Peritonitis (FIP)
Feline *Giardia*

yellowed, stained, tartar-encrusted teeth need your veterinarian's attention: a complete dental cleaning and scaling under anesthesia. Don't settle for a so-called anesthesia-free cleaning; it just won't do the job. (See box.) After a complete veterinary dental cleaning, you and your cat will enjoy a clean, healthy start for your at-home dental care program. And yes, that might mean brushing your cat's teeth.

Brushing Your Cat's Teeth

If possible, start brushing your cat's teeth in kittenhood. But even middle-age and elderly cats can learn to sit through regular cleaning—with time, practice, patience,

Don't Settle for Non-anesthesia "Dental Cleanings"

Many non-veterinarians, including groomers, are now offering "anesthesia-free dental cleaning" for cats. Some of these practitioners call themselves "pet dental hygienists" or "pet dental technicians." But no recognized veterinary regulatory body offers any training, licensing, or certification programs, or any oversight, for these service providers.

Most cat owners have a vague notion that regular dental care is important for their cats, but the differences between a cosmetic cleanup and a full dental cleaning performed by a veterinarian escapes them. The new teeth-cleaning practitioners often take advantage of this ignorance, implying that their service provides all the benefits of the veterinary procedure. This is something a lot of cat owners really want to hear. Why?

Many people are terrified of anesthesia, for themselves and for their pets. Many cat owners also find it difficult to understand why general anesthesia is necessary for something as "routine" as teeth cleaning. The short answer is that a thorough, effective cleaning is painful. A complete veterinary dental cleaning involves cleaning and scaling the teeth beneath the gum line, where most periodontal disease starts and festers. This procedure is painful and requires minimal movement of the patient, for both accuracy and safety. It's virtually impossible to accomplish this in cats without anesthesia. Even a complete veterinary dental examination—including beneath-the-gum probing and dental X-rays—is generally impossible unless the cat is anesthetized.

In addition, polishing the teeth after a thorough cleaning and scaling is essential to the continued health of teeth and gums, and polishing cannot be properly done without anesthesia. Polishing fixes and smoothes the invisible but potentially harmful damage to teeth caused by an effective cleaning and scaling. Cleaning leaves scratches and pits that can trap bacteria. Polishing fixes and smoothes out this damage. Without thorough polishing, the teeth and gums can be more susceptible to bacterial infection than before cleaning.

and a positive attitude. Start slowly, until your cat is accustomed to having his mouth and teeth touched. Then try rubbing his teeth and gums with a gauze pad or cotton-tipped swab moistened with a toothpaste or liquid made especially for cats. (See Resources.) My cats like the poultry flavor of C.E.T. Enzymatic toothpaste for dogs and cats so much they'll lick it right out of the tube. (But they still don't like brushing!) Don't use toothpaste or mouthwash made for humans; these products can irritate and burn your cat's mouth and stomach. You can even start out with plain water or the juice from a can of tuna.

Thorough polishing of kitty teeth takes time, patience, and meticulous care. An un-anesthetized cat just won't sit still for it.

In a non-anesthetic cleaning, tartar is removed only from the visible parts of your cat's teeth. Sometimes, to the extent that the cat permits it, the mouth is given a visual examination and the visible parts of the teeth are polished. Because the cat is awake, he may struggle and experience significant stress. Afterward, the cat's mouth looks and smells better, but nothing has been done about the real culprit in periodontal disease: buildup beneath the gum line. Non-anesthetic cleanings leave the owner with a false sense of security; a cat who gets regular anesthesia-free cleanings may be developing serious dental disease without the owner suspecting a thing until extensive damage has occurred.

Another essential benefit of a complete dental exam and cleaning under anesthesia is protection of your cat's lungs. Up-to-date veterinary dental protocol calls for use of a cuffed endotracheal tube that prevents the animal from inhaling tartar and other debris from his mouth during the cleaning. Inhalation of debris, always a possibility during an anesthesia-free procedure, can lead to pneumonia and other respiratory problems.

Ask lots of questions when scheduling a veterinary dental cleaning for your cat. Does your veterinarian use up-to-date anesthetic and pain-control protocols? Will there be a trained technician solely dedicated to monitoring your cat's heart rate, breathing, temperature, and other indicators, and making sure your cat stays warm enough? What about pain control? Sufficient pain control reduces the stress of anesthesia, allows swifter recovery, and can even lessen the amount of anesthetic needed during dental procedures.

Never hesitate to ask your veterinarian to refer you and your cat to a specialist, or to a veterinarian who is more experienced in providing dental care. Though there are fewer than one hundred board-certified veterinary dental specialists in the United States, there are many veterinarians who frequently treat dental issues and have the equipment, staff, and expertise to offer state-of-the-art dental care.

Hold your cat firmly, facing away from you. Hold his head with one hand and, with the other, slide the cotton swab or gauze into the corner of his mouth and gently rub his teeth and gums. Clean the side, rear, and front teeth. You don't need to clean the top or inner surfaces of the teeth. Thirty seconds of rubbing should be plenty, especially at first. Try for at least three cleanings a week. Every day is better. If this goes well for several weeks, experiment with a finger brush or tiny toothbrush made just for cats. Make dental care part of your daily grooming, petting, and home-checkup routine—and praise your cat lavishly for his forbearance.

If your cat turns out to be one of these "Absolutely not!" types, have your veterinarian take care of his routine dental care. For a thorough job, your cat will most likely have to be anesthetized. (See box on page 56.) In any case, discuss with your veterinarian the best approach to take with your cat. But no matter your cat's opinion on the matter, never ignore this vital, health-preserving care.

Ask your veterinarian whether a "dental diet" might be right for your cat. This dry kibble chow helps clean your cat's teeth and gums as he eats. The Bobcat Mountain cats have maintained excellent dental health despite their distaste for toothbrushing. They get Hill's Prescription Diet t/d (see Resources) dental diet (a complete, balanced diet in itself) as their treat food. They have great teeth. Since we've been using this food as daily treats, only a few of our cats have needed dental care under anesthesia.

MAINTAIN YOUR CAT AT A HEALTHY WEIGHT

Weight gain often sneaks up on middle-age cats (ages two through eight; see box on page 60), just as it does on middle-age humans. Like us, cats gain weight if they consume more calories than they burn. Combine the typical indoor cat's couch-potato lifestyle with boredom and highly palatable chow available around the clock, and you have the makings of a fat-cat epidemic.

According to 2010 figures from the Association for Pet Obesity Prevention, an estimated 54 percent of US cats are overweight (more than 10 percent over ideal body weight); about half of these are obese (more than 15 percent over ideal weight). Obesity is the number-one nutritional disease in cats—more common than all other nutritional diseases put together. This is not good news.

The Dangers of Obesity

A fat cat is neither a happy cat nor a healthy cat. For cats, as for humans, excess weight brings increased risk of disease; lessens the quality and enjoyment of life; affects mood, hygiene, and emotional health; and can seriously shorten life. Elderly cats are seldom obese; fat cats don't live that long. Overweight cats suffer more arthritis, bone and joint disorders, respiratory difficulties, asthma, heart disease, hypertension, pancreatitis, urinary tract disease, diabetes mellitus, stroke, constipation, chronic matting, and non-allergic skin conditions (such as feline acne and dry, flaky skin) than cats of healthier weight.

Fat cats run a much higher risk of dying in middle age than their slim cousins. Their extra weight heightens their risk of developing a serious liver disease, hepatic lipidosis, if they stop eating for more than a short time. They're more prone to sprains, pulled muscles, lameness, and other injuries caused by falls or jumps. Fat cats have trouble reaching all areas of their bodies to self-groom, so their hygiene and coat condition deteriorate. Unable to keep themselves clean, they can become depressed and even ill.

Overweight cats are at increased risk for complications from anesthesia. They take longer to recover from illness, injury, and surgery. Their immune systems can be impaired. Reduced energy levels, breathing difficulties, and extra strain on joints continue the vicious downward spiral. As leaping, running, and play become more difficult or painful, overweight cats become more reluctant to move at all, resulting in increased lethargy, apathy—and more weight gain.

Why Cats Get Fat

Barring certain medical conditions, the chief cause of feline obesity is us. Busy and preoccupied, we resort to cupboard love—overfeeding, dispensing tons of treats—and neglect to ensure our cats get the vigorous exercise they need to run off all those extra calories. Recent studies confirm that many cat owners are not aware that their overweight and obese cats are too fat, perhaps because they see so many other overweight cats—it looks normal to them!

When we bring cats indoors, treating them to a safe, comfortable, even luxurious lifestyle, we also remove them from their natural daily activities. A cat out on his own is a busy, active, constantly stimulated animal: patrolling and defending territory; maintaining guard against a variety of threats, from wild weather to predators; competing for resources; and stalking and killing prey in order to survive. Even while catnapping, he's alert to danger and opportunity. The necessity to track down and kill his meals helps ensure that his caloric intake remains balanced with his energy expenditure.

But when cats live exclusively indoors, all this changes drastically. This major lifestyle change places on us—instead of on the natural environment—the responsibility for providing safe, acceptable substitutes for natural activities and sources of daily stimulation so essential to maintaining feline mental, physical, and emotional well-being.

Although feline obesity may seem merely a matter of too much chow and too little exercise, there's also a strong psychological component. Excessive weight gain in many indoor cats can be chalked up to boredom, frustration, and depression. Mealtime becomes the only real pleasure in an understimulated cat's day.

Is Your Cat Too Fat?

If you think your cat is too fat, he probably is. Look down at your cat. Ideally, he should have a "waistline" (abdominal tuck) behind his ribs. Feel your cat's midsection. An overweight cat's ribs have a moderate covering of fat and are difficult to feel, and you'll see thickening at the base of his tail. In an obese cat, you'll feel prominent layers of fat over the ribs and tail base. If you can "pinch an inch" over your cat's ribs, he's too fat. Overweight cats get fat in their flank folds and have hanging folds that sway as they walk. Obese cats have drooping abdomens with

Yikes! Where did that paunch come from?

grossly thickened fat pads. And if you can't see your cat's elbow outline when he's standing—you guessed it—he's too fat.

Preventing Obesity

Outsmarting obesity is easy if you focus on prevention. Start early; weight problems can sneak up on cats as young as one or two years old, as the high energy needs of kittenhood diminish.

- Weigh your cat regularly. Keep records. Be alert for gains or losses.

- Watch your cat's everyday eating habits. If he has a tendency to snack non-stop, don't leave food out all the time ("ad libitum" feeding). Serve measured portions at set mealtimes, twice a day.

- Don't overfeed your kitten. Overfeeding during the rapid growth stages of kittenhood can cause an irreversible increase in the actual number of fat cells in the body. Overfed kittens may acquire hard-to-reverse habits of overeating, begging for treats, and seeking solace and stimulation in food.

- After your cat is spayed or neutered, cut back on his usual chow ration. Neutered male cats require about 28 percent fewer calories than unaltered males; spayed female cats require about 33 percent fewer calories than their unspayed sisters. That's because the higher estrogen and testosterone levels in unaltered cats help suppress appetite. If offered free-choice feeding, spayed females are four times as likely to be overweight than intact females, while neutered males are only twice as likely as intact males to get too fat.

- When you see your cat eyeing the empty food bowl between meals, refill his water bowl instead. Put some ice cubes in the bowl to make it more interesting.

- Don't misinterpret requests for attention and play as demands for food. Giving your cat treats or extra food instead of attention strengthens the link in his mind between the closeness he craves with you and extra food.

- Cut out the junk food. Many commercial cat treats are carbohydrate-laden junk. Serve healthy treats (see Resources) and make your cat work for them. Roll or toss treats so your cat has to jump or run for each one.

- If you just *have* to give in, put just a dozen or so kibbles in the bowl.

- Watch for competition at the food bowls. A dominant cat might take more than his share to establish and maintain his spot atop the pecking order, becoming overweight in the process. If you observe this, feed your cats in separate locations.

- As your cat approaches middle age (four to five years and older), watch his intake. As in people, energy requirements tend to decline in middle age, while consumption remains unchanged—a recipe for thunder thighs and potbellies.

- Engage your cat in vigorous interactive play every day. Ten to fifteen minutes, two or three times a day (right before mealtimes and at bedtime), is ideal. Provide plenty of exercising, climbing, and running opportunities for when you're not around to play athletic director.

Losing Weight Safely

If your tabby is too tubby, work with your veterinarian to formulate a plan and get started on a slow, safe, controlled weight loss and exercise regime—and stick with it!

- Schedule a veterinarian visit for a complete medical checkup to rule out underlying illnesses, and assess your cat's overall condition. Your veterinarian will assess your cat's "body condition score," based on his overall health, current weight, appearance, age, activity level, and other factors. Together, you and your veterinarian will decide on a weight loss goal for your cat and a sensible, healthy strategy to meet it.

  Your veterinarian will have a lot of questions about your cat's lifestyle and daily activity levels. Working with you, your veterinarian will set a weight loss goal for your cat and craft an effective weight loss strategy. Your veterinarian needs an accurate estimate of the cat's current intake—not just meals, but also treats, table scraps, any other food sources, and those "little extras." (Do the neighbor kids come over after school and slip your cat treats?)

- Feline weight loss involves a complete lifestyle change: diet, exercise, environmental enrichment, and increased daily interaction.

- Get a pet or infant scale, calibrated in ounces, to accurately track progress. The maximum safe weight loss rate is 1 percent of your cat's total body weight per week. In a 20-pound cat, that's just 3.2 ounces per week.

- Get the whole family on board. No diet will work if one family member keeps slipping your cat tuna on the sly.

- If your cat's just a little pudgy, serving slightly smaller portions of her usual chow, cutting out the treats, and stepping up the exercise might do the trick.

- If your cat's seriously fat, be prepared for a lengthy project: several months to a year. Your veterinarian will likely prescribe a switch to a special diet food. Switch over gradually, over a one- to two-week period. Think of the prescribed food as medicine. Stick closely to recommended portion sizes and feeding schedule.

- Be sure your veterinarian knows about and approves all supplements, vitamins, remedies, or other additives you're feeding your dieting cat. These can affect the overall nutrient balance.

- No matter how fat he is, *never* put your cat on a crash diet. This could bring on hepatic lipidosis, an often fatal liver disease. Never let your cat go without food longer than a couple of meals at most.

- Some cats will try hunger strikes. If your cat completely refuses the diet food, ask your veterinarian to recommend another flavor or brand. Never let your cat starve. Missing just a few meals can be very dangerous.

- Resist begging. Go cold turkey on "kitty junk food" and table scraps. If your cat's a particularly persistent or heart-wrenching beggar, lock him out of your bedroom at night, or keep him away from the dinner table during meals.

- If you feel you must give your cat treats, serve healthy treats (see Resources) or use a few morsels from his daily measured allocation of diet food.

- Take a hard look at your cat's environment. Does he have plenty of opportunities for exploring, climbing, playing, and entertaining himself without resorting to nervous snacking? Might a companion cat or kitten put some verve, interest, and daily stimulation back into his routine-bound life? Often the best exercise for a cat is another cat.

- Step up the exercise. Start slowly and be persistent. Avoid overexertion, especially at first. Don't make your cat exercise if it's very hot or humid, or right after eating. And don't expect miracles. If your cat is unused to anything more strenuous than a slow waddle to the food bowl, even a few roll-overs are enough for a start.

- Work up to several-minute sessions of interactive play with a fishing-pole toy such as "Da Bird." (See Resources.) Make playtime exciting, fun, and special so that your cat will come to look forward to it. Give plenty of praise and reinforcement. Show your cat you're having fun, too! Quit when he starts losing interest, preferably before he's had quite enough. And always stop immediately if your cat shows signs of exhaustion, fatigue, or shortness of breath.

- Check in with your veterinarian at least monthly for weigh-ins and an assessment of overall progress.

- Weight loss is a tough, long-term project. When you get discouraged, remind yourself of the serious health consequences of obesity.

- When your cat reaches his weight loss goal, your veterinarian can recommend a lifelong diet program to maintain optimal weight. This might include portion-controlled feeding of a weight-maintenance food.

- Stay on guard. Don't slip back into old bad habits of too many treats and a couch-potato existence.

Once he drops about the first third of his excess weight, your cat will start feeling happier and healthier. He'll have more energy and be a livelier and more active, playful, and attractive companion.

YOUR CAT'S WEEKLY AT-HOME CHECKUP

About once a week, give your cat a good nose-to-toes once-over. Look for signs of pain or illness, and for any changes in appearance, energy level, habits, movement, or behavior. If you observe anything out of the ordinary, a visit to the veterinarian may be in order.

First, weigh your cat. If possible, use a scale calibrated in ounces so you can detect relatively small changes. Write down his weight in a notebook or on your calendar so that you can keep track of changes. (I write down kitty weights on a large calendar that hangs on my refrigerator door.) A loss or gain of a half-pound or more is cause for concern, especially if the gain or loss is rapid. Rapid weight loss is often the first sign of serious underlying illness. Older cats in particular can sicken extremely quickly.

Check your cat's nose. Cats sneeze and snort occasionally, just like we do. But a constantly sneezing cat may have an upper respiratory infection. Other signs of respiratory problems include discharge from the nose or eyes, sniffling, and ragged or wheezy breathing.

Look into your cat's eyes. They should be clear, with no bulging, squinting, cloudiness, or redness. Look for the "third eyelid" (the white fold of skin that stretches from the lower eyelid to the inner corner of your cat's eye). You should see only a tiny sliver. If you see a lot of white showing in this "nictitating membrane" when your cat is fully awake, it may be a sign of illness.

Sniff your cat's ears. A yeasty or sour odor may mean an ear infection. Try to get a good look into his ears with a small flashlight. Look for crusty, brownish or black discharge; red, irritated ear canals; and any signs that your cat has been scratching or clawing at his ears. This could mean ear mites. Ear mites are highly contagious; if one cat has them, other cats in your household probably do, too.

Run your hands gently all over your cat's body, from his head to the tip of his tail. Can you feel his ribs? You should be able to. If not, he may be too fat. Does your cat cry, flinch, or vocalize when he's touched in certain spots? Make a note of it. Check all over for swelling, bumps, and lumps.

Run your hands along your cat's underside, paying special attention to his belly. Is he reluctant to be touched there? Do you see patches of missing fur? Cats with abdominal pain often tear out tufts of fur from their bellies. Does his abdomen seem puffy or distended? Does he seem to have a "potbelly"? Are there scaly patches, lumps, or bumps, especially near his nipples?

The overall condition of your cat's coat reflects his diet, health, and emotional state. If he's ill or depressed, his coat and his grooming will suffer. If he's eating poorly or eating low-quality food, his coat will reflect this. An ill cat will often have a dry coat, scaly skin, and missing patches of fur, or he may shed excessively.

How's his grooming? Grooming is a basic feline instinct. Cats who are depressed or ill often stop grooming themselves, a clear danger sign of illness.

If your cat has neglected grooming his rear end, it may be because he's overweight and can't reach that far anymore. Or he may be suffering from stiff joints or arthritis. Elderly cats in particular appreciate some gentle help with their grooming. Use a warm washrag, soft hairbrush (try a baby's brush), and a light touch. If your longhaired cat has extra-long bloomers and is having trouble keeping them tidy, trim back his fur a bit or see a professional groomer.

Cats who venture outdoors, even if only on a harness and leash or into an outdoor enclosure, may be exposed to fleas, ticks and other parasites, and biting insects. But even if your cat is an indoor-only pet, it's a good idea to check regularly for any signs of external parasites.

Is your cat itchy? Does he paw, scratch, or dig at his skin? Is he pulling out tufts of fur? Do you see scabby or reddish raised patches? Your cat may have not only fleas, but an allergy to flea bites: flea allergy dermatitis, also called *miliary dermatitis.*

Does your cat have roundish, scaly, reddish patches of missing fur, especially around his face, muzzle, and ears? He may have ringworm. Ringworm isn't a worm at all, but a fungal infection communicable to other cats and to humans. (See chapter 13.) Because of cats' face-to-face-rubbing greetings, the fungal spores are often transmitted to other cats' faces, ears, and front legs first.

If you see anything suspicious during your weekly checkup, write down a few notes and schedule a visit to the veterinarian.

PAY ATTENTION EVERY DAY

Cats are notoriously stoic and they are *extremely* good at hiding illness. Millions of years of evolution as solitary predators has left felines extremely reluctant to betray any sign that they are ill or disabled in any way, for fear that a competitor for their all-important territory might take advantage. Even though this highly competitive, life-or-death environment doesn't prevail in our households, our cats are still so close to their wild cousins in behavior and outlook that it can be devilishly hard to tell when they're not feeling well. So your cat depends on you to notice signs of illness in his behavior and daily activities.

Even if you don't suspect an illness, always keep a close watch on your cat's everyday habits and activities. By knowing what's normal for your cat, you'll be much more likely to recognize changes that could signal illness or pain.

Sometimes you might just have a vague feeling that "something's not right" with your cat. Pay attention. After all, you know your cat better than anyone. If your cat just seems "off," consider taking him in for a checkup.

Pay attention to your cat's normal gait, movements, posture, and energy level and watch for changes. For example: Does he seem stiffer than usual, especially in his spine or neck? Does he exhibit pain or obviously limited movement? Is his balance off? Is he favoring a leg? Does he seem overly sleepy, lethargic, or not as interested in life as he usually is? Is he skipping playtime? All these can signal illnesses and warrant a visit to the veterinarian.

The Telltale Signs of Feline Illness

As a general rule, *any* major or sudden change in your cat's usual activities and habit patterns calls for vigilance and investigation. Don't ignore these signs of potential illness:

1. *Refusal to eat or eating a lot less than usual (for example, taking a bite or two, then walking away and not returning), for more than a couple of meals:* Food refusal or cutback always needs close attention, as it can point to so many different problems, some of them quite serious. (See chapter 8.) If the refusal to eat continues for more than a few meals, the cat may be at risk for developing *hepatic lipidosis*, a serious liver disease. In a cat, persistent refusal to eat is *always* serious business.

2. *Eating only on one side of the mouth, refusal to eat hard or crunchy food, dropping food bits out of mouth, vocalizing while eating:* This might mean painful dental problems, such as FORLs (cervical line lesions). FORLs need veterinary attention as soon as possible, as the severe pain can cause a cat to stop eating entirely.

3. *Hiding and withdrawing from contacts:* Has your usually friendly, outgoing cat taken to hiding in closets? Is he avoiding contact with other cats and people, or skipping once-favorite activities such as playtime? This behavior may signal pain or illness, especially if these changes come on suddenly.

4. *Unexplained weight changes, especially if rapid:* A pound or two for a cat is the equivalent of ten or twenty pounds for us. Cancer, hyperthyroidism, and kidney and liver disease, among others, can cause rapid gain or loss. These are diseases you want to catch as early as possible.

5. *Lashing out or unusual aggression.*

6. *Getting tired extremely easily, wheezing after just a few jumps, losing interest quickly in games he used to enjoy:* This could be "exercise intolerance," which, especially in a kitten or young cat, can be a sign of heart disease.

7. *Any change in elimination habits or patterns:* Is he pooping or peeing outside the litter box? Spraying the walls? Leaving puddles of urine on the floor? Dousing the carpets? Don't assume your cat is mad at you, jealous of your new boyfriend, or just being naughty for no good reason. Lapses in litter box etiquette can be caused by either medical or behavioral issues (or sometimes both) that can be difficult to sort out. (See chapter 5.) But the first step in *all* cases of "outside the box" thinking is a trip to the veterinarian for evaluation and testing.

8. *Drinking a lot more water than usual, and/or urinating extra-large amounts:* Empty water bowls and huge clumps in the litter box can indicate that the cat's kidneys are not filtering toxins efficiently. Drinking and urinating more copiously than usual are often the first signs of chronic kidney disease, a condition much easier to manage if caught early.

9. *Flinching from touches, such as stroking or grooming:* This can signal pain or injury, especially if the cat ordinarily enjoys these kinds of contact. Note which areas seem to be the most sensitive so that you can tell your veterinarian.

10. *Changes in vocalization:* A silent cat who suddenly cries and calls at all hours, or a talky cat who's suddenly mute, may be experiencing pain, stress, or confusion.

11. *Constant head-shaking, violent digging, and scratching at one or both ears:* This can indicate ear mites (a parasite), or a bacterial or fungal infection. A veterinarian can tell the difference, but don't try to diagnose this yourself. If you leave ear mites untreated in one cat, soon every cat in your household will have them—if they don't already.

12. *Crouching in a corner with head facing into the corner:* This is an odd one, but it is usually a sign that the cat is in severe pain or feeling extremely ill.

13. *Persistent bad breath:* Persistent bad breath can point to a variety of illnesses, including periodontal disease, kidney failure, and gastrointestinal problems. It's not just a cosmetic issue.

14. *Change in grooming habits:* Grooming is a basic, essential feline activity. If a cat stops grooming himself, or even reduces the amount of grooming that's normal for him, pay attention. Regular self-grooming has a positive effect on a cat's overall health, mood, and morale, and lack of grooming can be an early sign of pain, illness, or depression. Keep an especially close eye on elderly and overweight cats, who often have trouble grooming themselves thoroughly.

Look to the Litter Box for Clues

The first signs of illness often appear in the litter box. That's just one reason for keeping your cat's litter boxes in an easily accessible, visible spot. Get familiar with your cat's normal output patterns, and you'll be more likely to notice any changes.

Here are some things to be on the lookout for:

1. Empty litter box: There are two possibilities:

    If your cat is eating and acting normally, but nothing is appearing in his litter box, he may be "thinking outside the box." Hunt around the house with a flashlight in corners, closets, and other out-of-the-way spots to find out where your cat has been going. Then see chapter 5 for what to do about his unfortunate lapses in litter box etiquette.

    If your cat appears to be eating but is producing no urine or feces that you can find, he may be constipated, obstipated, or have a urinary blockage. He needs immediate veterinary attention.

2. *Blood in urine or feces:* While not necessarily an emergency, this always calls for a prompt visit to the veterinarian.

3. *Worm segments in feces:* These can look like strands of long, thin spaghetti or grains of rice. Place a fresh feces sample in a sealed plastic bag and take it to your veterinarian for testing. Don't just give your cat a generic or over-the-counter de-wormer. The specific type of worm, as well as your cat's age and medical history, can call for different remedies and medications.

4. *More or less urine than usual:* Keep an eye on the size of the urine clumps or puddles. Much larger than usual clumps, or several tiny ones, could be a sign of illness. In male cats, urinary blockages can be fatal within hours. *If your cat strains in the box but produces no urine, or just tiny drops, it's an emergency.* (See chapter 14.)

5. *Unusual appearance of feces:* Be aware of the shape, size, color, and amount of feces your cat usually produces. Pencil-thin feces can indicate a partial blockage. An unusually large amount of hair in the feces may mean your cat is self-grooming more than usual due to stress, pain, or itching.

6. *Diarrhea:* Unless it's gone in twenty-four hours, diarrhea requires at least a phone call to the veterinarian. Have you changed brands of cat food? Has your cat eaten something unusual? Milk gives many cats diarrhea. (See chapter 5 for cleanup tips.)

7. *Unusually strong odor:* Your cat may have eaten something unusual. If he's "scooting" on the carpet or licking and grooming his anal area excessively, his anal glands may be full or partially impacted, causing both his distress and the foul odor. The situation usually resolves itself within a day or so. Otherwise, a trip to the clinic is in order so that the cat's anal glands can be emptied or "expressed." Most cats never experience this problem, but for some it's a chronic issue. You can learn to express your cat's anal glands at home, if both you and your cat are willing.

If you have multiple cats sharing litter boxes, try keeping watch on the boxes shortly after mealtimes (when most cats make their visits) and see if you can learn which cat is leaving the unusual urine or feces. If you've already noticed a particular cat showing other signs of illness, isolate that cat—with his own litter box, cleaned first—in his own room for further observation.

Outsmarting Medication-resistant Cats

Many cat owners consider administering medications to cats to be one of their biggest challenges. Hiding medications in food, the old trick that works so well with dogs, is a bad idea with cats for a couple of reasons:

1. It generally doesn't work. Cats are extremely sensitive to the smells and tastes of their food. If anything seems slightly different or "off," the feline tendency is to refuse it.

2. It tends to make the cat suspicious of anything that appears in his food bowl. A cat who's ill enough to need medicine will be at increased risk if he decides to stop eating because he's not sure what kind of suspicious foreign matter you'll try to slip into his food.

MEDICATION OPTIONS

What does work? The first thing to understand is that you and your cat have options. Many medications come in alternative forms, including cat-appealing flavorings, transdermal creams and gels, and flavored powders that can be mixed with foods. Your veterinarian will be happy to work with you to find a form of the needed drug that will suit you and your cat. Most veterinarians work closely with compounding pharmacies, specialists who are wizards at transmuting yucky liquids into tasty ones, hiding the taste of bitter pills, or avoiding pills altogether.

If your veterinarian prescribes a liquid medicine, but you know your cat accepts pills better than liquids, ask if the medicine comes in pill form and vice versa. Many drugs come in both forms. Are the pills large or small? Capsules or tablets? Can pills be broken in half for easier dispensing? Can capsule contents be removed and mixed with, say, tuna juice? The answers to these and similar questions can depend on the individual drug being prescribed. Don't assume— ask! And be sure to ask whether the taste of the prescribed drug is known to be bitter or unpleasant to cats. Having this essential information will help you be prepared for your cat's reaction and motivate you to secure a handy supply of tuna juice or tasty treats for medication time.

If you know that your cat is violently opposed to all medications given by mouth, or you're just tired of getting bitten, ask whether you can use a transdermal cream or even give the drug by injection. In extreme resistance cases, or

A Spoonful of Water Helps the Medicine Go Down

Administering water right after medicine makes the process easier on you and your cat, and it can also prevent complications. A study funded by the Winn Feline Foundation ("The Evaluation of the Esophageal Transit of Tablets and Capsules in 30 Cats"), performed at Colorado State University's College of Veterinary Medicine & Biomedical Sciences, looked at how long pills can remain in a cat's esophagus. The investigators were particularly concerned about the commonly used antibiotic doxycycline. They discovered that doxycycline tablets stayed in the esophagi of the cats long enough to cause severe inflammation (esophagitis) and resulting scars.

"The investigators concluded that administering 6.0 ml of water after tablets or capsules will greatly decrease the risk of esophagitis and stricture formation, and ensure that all of the medication enters the stomach which is essential for adequate dosing. Most household teaspoons hold about 5.0 ml of water. Therefore, it may be prudent for veterinarians to instruct owners to administer water to their cat after giving any tablet or capsule, but especially doxycycline."

But even if the pills your cat needs won't cause such severe problems, imagine how you feel when you try to swallow a pill with too little water. Not pleasant, right? The administration of water right after medication works almost like magic to make the whole experience quick, easy, and painless for your cat—and you. It just takes a little practice.

when administration of a drug is more complicated or delicate than you want to attempt at home, it might be better to just bring your cat to the clinic for his daily medication. There are lots of options, both for the good of your cat's health and to maintain your sanity. But you won't know about the options unless you ask.

Get It Right

Before you leave the veterinary clinic with your cat and a bottle of pills or liquid medication, be completely sure that you know exactly how much of the medication to give, how often to give it, when to give it, and what effects, side effects, or signs of trouble you should be on the lookout for. Some of this information will be printed on the label, but it's still best to ask. Don't leave the clinic until you're sure you understand the instructions. Bring a notebook to write down the details, or even ask the doctor to repeat the medication instructions into a portable voice-recording device.

If the label says "Give 1 ml," make sure you know exactly how much that is. Get out the dispensing item (such as the dropper or oral syringe) and ask the doctor or a technician to verify where the "1 ml" line is. But what if the syringe is labeled for "cc's"? How many cc's is 1 ml? Ask! Don't worry that you'll sound stupid; the only stupid questions are the ones you don't ask.

Still confused or have additional questions about the medication after you get home? Call the clinic and ask! Your veterinarian and the clinic staff would much rather answer your questions three times than see your cat go without his medication.

No matter your cat's opinions on the subject of medications, follow your veterinarian's dosage and duration directions exactly. Don't stop medication just because your cat seems to feel better. Stopping a course of antibiotics too soon is particularly foolish, as it encourages chronic reinfections and the development of antibiotic-resistant organisms. Many cats suffer from chronic or recurrent urinary tract infections because the first infection was incompletely treated.

Okay, so you're home with your bottle of pills and your cat is eyeing you warily. What are the best strategies and techniques for pet medication success?

HOW TO PILL A CAT

Over the years I've learned techniques and strategies that, while not guaranteed to succeed every time with every cat, at least maximize the chances that

1. the cat will actually get the medicine;

2. the cat will not be so traumatized by the experience that future medication attempts will be even more difficult; and

3. the human will lose as few body parts as possible.

Medicating a cat calls for speed, dexterity, agility, confidence, and grace. It also calls for a cat. Cats (and owners) for whom pill time has not been a pleasant experience are understandably reluctant to repeat it, and cats have a mysterious way of vanishing at the first hint of a pill or syringe. Remind yourself:

1. how important it is for the medicine to be able to do its healing work in your cat, and

2. that the medicine won't help if it stays in the bottle or gets spit out behind the sofa.

Your first task is attitude adjustment. You *will* succeed. Stay calm and focused, talk softly to your cat the whole time, be determined yet gentle, and be sensitive to your cat's reactions and feelings. Once you've mastered a few key tricks and techniques, pilling cats gets easier. No, really! My calico Chrysanthemum used to race away from me in panic, foam at the mouth, and growl if she even suspected I might have a pill in my hand. These days, though, she sits calmly after breakfast waiting for her daily medicine and takes it peacefully.

Next, prepare the pill or other medication so it's fully ready to administer. If a liquid needs to be placed into a dropper or oral syringe (I highly recommend the latter), get it ready and then re-cap the bottle. If it's a pill, remove it from the bottle and re-cap the bottle. Also, and this is key, prepare a second oral syringe of plain water or tuna juice–flavored water.

If the cat is sitting on the floor, casually lower yourself to the floor behind him. If he races away, calmly follow him and pick him up when he lets you. If he seems very resistant, don't insist. Wait awhile and try again. A word of warning: NEVER grab your cat to medicate him while he's eating, sleeping, or using the litter box. For one thing, it's rude. But it can also backfire badly, causing him to avoid his litter box, lose his appetite, or fear your every approach.

Ready? Remember to talk calmly and gently to your cat during this entire procedure. Explain what you're doing. Use the same words every time so he knows exactly what to expect.

1. Kneel on the floor behind your cat.

2. Slither forward while backing the cat up between your legs so he's facing away from you. This makes it tough for him to wiggle away, because you're controlling most of the perimeter (if you think in terms of military strategy).

3. With one hand, tilt your cat's head up a bit until his lower jaw drops down. (This is where your speed and grace come in.)

4. If you're giving a pill, you may have to quickly pry open his mouth a bit wider so that you can see down his throat, like looking down a tunnel. (You might have read or heard that coating a pill with butter might make it go down easier. In my experience, though, it just makes the pill slippery and hard to handle.)

5. Pop the pill right down that tunnel, down the middle of the tongue, past the "hump" in the tongue. It's important to get the pill right down

How Old Is Your Cat?

Cat Age	=	Equivalent Human Age
6 months	=	10 years
8 months	=	13 years
1 year	=	15 years
2 years	=	24 years
4 years	=	32 years
6 years	=	40 years
8 years	=	48 years
10 years	=	56 years
12 years	=	64 years
14 years	=	72 years
16 years	=	80 years
18 years	=	88 years
20 years	=	96 years
21 years	=	100 years

the middle, or it'll get stuck in a "side pocket" from which it will inevitably get spit out. If you miss getting the pill right down the middle, let the cat spit it out (as he'll most likely do) and start over.

6. Squirt in some water from your water syringe immediately. This helps the cat swallow the pill quickly and easily. If he's having trouble getting the pill down, offer more water. Squirt water from the side of the cat's mouth. Don't aim directly at his throat. The idea is to help the pill slide down, not drown the cat.

7. If you're giving liquid medication, squirt it similarly, from the side. If it's a large amount, try giving half first, offer some water, then give the rest. This is especially helpful for sticky or bad-tasting medicines.

8. Once you're sure he's swallowed the medication and he seems comfortable, offer lavish praise and a little kiss on his nose.

Chapter 4

The Secrets of Your Cat's Mind

What's Important to Your Cat?

"We cannot without becoming cats, perfectly understand the cat mind."

—*St. George Mivart,* The Cat *(1900 edition)*

To successfully outsmart your cat, learn how she thinks, how she sees the world, and what's important to her. Understanding your cat's behavior is much simpler once you know these Top 20 Secrets of Your Cat's Mind:

1. Your cat is a *cat*. She's not a "little person in furs." She's driven by cat motivations, cat urges, cat needs, cat concerns—not human ones. We humans and our cats have much in common, but we see the world quite differently in many ways. Blissfully unaware of the complex layers of meaning we humans tend to assign to everything we see and do, your cat is a supremely practical creature. Everything she does, she does for a reason. A perfectly good reason. A perfectly good *cat* reason.

   "With the qualities of cleanliness, discretion, affection, patience, dignity, and courage that cats have, how many of us, I ask you, would be capable of being cats?" —*Fernand Mery*

2. Your cat is a hunter. Her predatory nature is woven deeply and inextricably into her body, mind, and spirit. This can't be changed, ignored, or wished away. You can no more teach your cat not to hunt than you can teach her not to breathe. But you *can* channel her predatory drive into nonlethal but cat-satisfying activities that fit harmoniously into your life and home.

   "Prowling his own quiet backyard or asleep by the fire, he is still only a whisker away from the wilds." —*Jean Burden*

3. Your cat is an individual, unique in all the world. Adjust your expectations accordingly, and always be ready for surprises.

> "Cats are absolute individuals, with their own ideas about everything, including the people they own." —*John Dingman*

4. Your cat is scent-oriented. Subtle scents carry messages to and from her that we can't begin to imagine. She recognizes individuals and their intents, status, and agendas primarily by scent. Scents we humans can detect (urine, feces) as well as those we can't (facial pheromones) are vivid intercat communication media.

5. Your cat doesn't understand the concept of "command." She *may* take a suggestion, if it looks like there's something in it for her.

6. Your cat's an opportunist, always on the prowl for the best deal. Her opportunism is a vital survival skill. She doesn't wait for opportunity to knock—she nudges the door open herself.

Sharp eyes scanning, ultrasonic ears swiveling, whiskers fanning, paw pads tapping, she's ever alert for threats and opportunities. Dozing on the sofa, looking like she's tuned into outer-space-kitty-radio, she's actually taking in and processing input from her immediate environment. At the first hint of an interesting odor or sound, she's back on terra firma, poised to take instant advantage.

Living in the moment, she makes decisions instantly, based on instant feedback from her environment. She doesn't hesitate. She makes no moral or ethical judgments. She can be bribed. *What's in it for me?* she wants to know. *Right now?*

> "To understand a cat, you must realize that he has his own gifts, his own viewpoint, even his own morality."
> —*Lilian Jackson Braun*

7. Your cat has enormous patience for things that interest her and no patience whatsoever for things that don't.

8. Your cat is a creature of habit. She craves routine, repetition, and ritual. She dislikes and distrusts change of any kind, especially if it's not her idea. She's happiest in a peaceful, harmonious, predictable environment, over which she feels she has a large measure of control. But she also reserves the right to change her mind.

9. Your cat craves stimulation, excitement, and activity, but in cat-manageable, cat-controllable doses. She wants *her* paw on the remote control at all times.

10. Your cat is territorial. To a cat, even a pampered indoor cat, territory equals survival. Carving out a territory and defending its resources from potential competitors is a fundamental cat concern, honed by millennia of evolution. It doesn't matter that her territory is your home, rather than a tract of forest. It's *her space.* She demands, and expects, those with whom she shares her life to respect that space. When she wants to be alone, she expects to be left alone.

11. Your cat lives, moves, and thinks in three dimensions. Because she's small and close to the ground, easy access to height provides her crucial psychological and physical advantages. Heights give her confidence and a sense of control over her environment.

12. Your cat is a highly touch-sensitive, affiliative animal who delights in communication, contact, interaction, and companionship—*on her terms.* She's also extremely intuitive, tuned in closely to human emotions. She always knows how you're feeling.

> "If you want to be a psychological novelist and write about human beings, the best thing you can do is keep a pair of cats."
> —*Aldous Huxley*

13. Your cat needs to feel that she's master of her own destiny. She needs to know she can make a swift, voluntary escape from any situation that frightens or displeases her. In times of danger and uncertainty, her first impulse is to flee or hide. Only if flight is impossible will she steel herself and prepare to fight.

14. Your cat is a polite, sensitive creature, and expects and deserves politeness and sensitivity from you and her other companions.

> "To respect the cat is the beginning of the aesthetic sense."
> —*Erasmus Darwin*

15. Your cat is a stimulus-driven, hyperalert, highly tuned animal who's unusually susceptible to both stress and boredom. She can suffer from stress when she becomes overstimulated by elements in her environment that she can neither control nor escape. She risks boredom when

How Long Does a Cat Remember?

Researchers at the University of Alberta in Canada wanted to find out whether cats remembered their own actions longer than things they only see.

To study how long cats retain these different kinds of memories, the researchers placed an obstacle in front of a cat, then stopped the cat's motion just after her front legs had crossed over the hurdle, but before her hind legs had crossed over. Then, while distracting the cat with food, they removed the obstacle. It turned out that the test cats remembered the presence of the obstacle for about ten minutes and kept raising their hind legs to avoid it—even though it was no longer there. But if a cat's motion was stopped just before she crossed the obstacle, she remembered its presence for only a few seconds.

So, to form a long-lasting memory, a cat needs to actually perform an action, not just see something. Remember this when training your cat and trying to get her to do what you want. Reward her after she does what you want her to do, and she'll remember it longer.

her natural urges for mental and physical stimulation, emotional connection, and especially her natural predatory urges are thwarted.

16. Your cat does whatever she does because it *works* for her. She chooses a behavior or habit pattern because it solves a problem for her in the most satisfying and acceptable way she can devise at that moment. Her solution is not always what you would choose. If you don't like her solution, it's up to you to outsmart her. Offer a superior alternative, a deal she can't refuse.

> "The best time to determine what you want from your cat is before he starts doing what you do not want." —*Phil Maggitti*

17. Your cat's misbehavior is often communication in disguise. She needs to get your attention and break through your preoccupations to convey information important to her. If you ignore or misinterpret these attempts at communication, she'll likely escalate her attempts until you *do* get it.

> "Cats are mysterious folk. There is more passing in their minds than we are aware of." —*Sir Walter Scott*

18. Your cat lives in the present. She does have a keen memory of people, places, and events. But she neither mopes about the past nor frets about

the future. She enjoys a rich emotional life and has a complex, intriguing mind, but she's eternally oriented to *now*. She focuses on solving the problem or savoring the joy of the present instant. Punishment, revenge, spite, evil, next week, and later have little meaning for her.

> "Unlike us, cats never outgrow their delight in cat capacities, nor do they settle finally for limitations. Cats, I think, live out their lives fulfilling their expectations." —*Irving Townsend*

19. Your cat is an adaptable, flexible, curious, and intelligent animal who can be trained, retrained, and outsmarted.

20. Your cat is a generous and clever teacher. She'll reveal everything you need to know to outsmart her—*if you pay attention*. When in doubt, look to her for guidance. Observe and note her individual quirks, habits, likes, dislikes, and personality traits: information you need to outsmart her.

> "I have studied many philosophers and many cats. The wisdom of cats is infinitely superior." —*Hippolyte Taine*

Domestic vs. Wild

On the leisurely time scale of evolution, small cats are very recent additions to human circles. Observe the similarities in size, structure, and behavior of virtually all domestic cats, regardless of breed. What's more, all domestic cats are remarkably similar in many ways to their closest wild cousins, the thirty-plus species of small wildcats, such as bobcats, ocelots, and margays.

Dogs and their ancestors have lived among humans for almost fifteen thousand years. Our early forebears were quick to recognize the usefulness of dogs and to preferentially propagate canine traits most useful to humans. Huge, powerful dogs helped hunters bring down big game, while small, wiry dogs ran rabbits and snakes to ground. Dogs that specialized in tracking, scouting, searching, hunting, protection, battle, herding, retrieving, and home companionship gradually emerged, varying greatly in size, shape, build, appearance, talents, and temperament. Domestic dogs have been so manipulated and changed by humankind that today's dogs are an entire genus—not just a species like domestic cats—removed from their closest wild relatives.

Cat Names and Cat Behavior

For nearly twenty years, I lived with a splendid calico, Petunia, undisputed queen of the household. Because of her exalted role and superior demeanor, I started calling her The Boss, or Boss Lady. She soon started throwing her weight around: hissing and swatting at lower-status cats, chasing them away from preferred spots even when she wasn't using them. Her new nicknames seemed to encourage a subtle but definite change in how she managed intercat relationships in her domain.

After watching her chase shy Dominique from the litter box for the third time, I decided enough was enough. On a hunch, I stopped calling her The Boss and started calling her Mama Cat, or just Mama. Amazingly, she toned down her bossy behavior, shifting gradually back to a more laid-back, maternal role. Intercat stress went down.

Did Petunia know what the words *boss* and *mama* meant? Probably not (though with cats, I'm never too sure). More likely, she responded to the subtle differences in my tone and inflection when I spoke those words. Perhaps *she* didn't know what they meant, but *I* did. My awareness of the meanings and differing implications of her nicknames came through to her loud and clear.

Go ahead and call your cat Lard-Butt, or The Monster. She won't "know" those words and won't be insulted by their literal meanings. But she will pick up your subtle inflections and perhaps try her best to live up (or down) to her nickname.

To a wildcat, a minute variation in the tone or volume of a sound could mean the difference between catching that mouse or not, or escaping from that predator or not. To a cat—details matter.

THE MYSTERY

If your cat is so close to the wild as to be nearly indistinguishable from her wild cousins, what gives her the uncanny ability to slip so comfortably and seamlessly into your life and home? Most adult small wildcats are fierce, hard to handle, unfriendly, and unpredictable, even if bottle-fed and hand-raised by humans. Anecdotal evidence abounds that keeping even a hand-raised wildcat as a pet is usually a recipe for disaster.

NEOTENY—THE MYSTERY SOLVED

Neoteny is the retention of juvenile features and behaviors in an adult animal. It explains why your cat is, in so many ways, a "permanent kitten." In the early

1960s, Russian biologist D. K. Bely-aev was breeding silver foxes, trying to make it easier for fur farmers to raise and handle the difficult-animals. He discovered that by simply selecting for a single characteristic—friendliness toward humans—he was able to speed up evolution. In a very short time, he bred what appeared to be an entirely new kind of fox. His new, *neotenized* foxes were as tame and approachable as domestic dogs. They also had many physical characteristics and behaviors common in young domestic dogs, but previously unknown in adult silver foxes.

When the small wildcats of the Fertile Crescent (*Felis silves-*

Wild or domestic, a cat is a cat is a cat.

tris libyca) decided to live among humans, it was likely the tamest, friendliest, most adventurous, playful cats—the most kitten-like—who made the first move. Besides their usefulness as rodent killers, it was their kittenish qualities—sociability, willingness to be petted and groomed, love of play, willingness to mingle in groups, and adaptability to human lives and homes—that so endeared these first cats to our ancient ancestors. The tamest, friendliest, most agreeable cats hung around and reproduced, eventually giving rise to a new species: *F. s. catus,* the domestic cat. Their wilder, warier cousins slunk back into the bush.

One of the nicest traits retained by our neotenized cats is their ability to regard humans and other animals as companions, or "honorary cats," rather than enemies, competitors, predators, or prey. (Tempting your cat with a gerbil friend or mouse buddy is still probably an unwise idea, though.)

Wild felines are high-strung, sensitive, finely tuned, territorial, solitary hunters, hyperalert to potential opportunities, threats, and competition. In your cat, these characteristics are still present, but they're smoothed out, damped, and calmed by the overriding effects of a lifelong juvenile mind-set and behavioral blueprint, mellowed by the power of neoteny. Knowing that your cat remains driven by deeply held wildcat instincts is essential in understanding and outsmarting her.

Tabigail the domestic cat and Kissy the mountain lion have a lot in common.

Our cats are no longer truly wild. But they're still not fully domesticated, either. Your cat, when properly socialized, is a tame, gentle creature whose dignified nature and relatively predictable behavior make her an ideal household companion. Yet she possesses an essentially wild soul. Relish her versatility, gentleness, playfulness, adaptability, and capacity for forming loving bonds. But also thrill to hints of that ever-present wild creature—her inner wildcat—in all its glory, mystery, and paradox.

So Alike, Yet So Different

In many ways, we and our cats are very alike.

- We crave a clean, safe, pleasant, harmonious, relatively predictable (but not too predictable!) environment.

- We like to have our own way.

- We crave novelty, fun, stimulation, and excitement—but in manageable, non-threatening doses.

- We want our fingers on the button, our hands at the controls, at all times.

- We enjoy conversation and touching, contact and closeness—but on our own terms.

- We get prickly when somebody invades our "personal space" without asking.

- We're sensitive to stress and resistant to change in our familiar daily routines.

- We love to play, to be entertained, to be the center of attention when we feel like it—and to be left alone when we feel like it.

- We love comfort, luxury, convenience, and being waited on.

- We thrive on fresh air, naps, warmth, and sunshine.

- We don't like being forced, coerced, threatened, manhandled, or tricked.

- We generally dislike surprises.

- Once we find our comfort zones, we want to stay there.

But we're also vastly different from our cats.

> "Well, I've lived with cats most of my life, so I'm very aware that there's another world going on. It's sometimes sitting in your lap, so obviously it's not completely different. But it sees everything differently, hears everything differently and probably thinks differently."
> —*Edward Gorey, cat-loving writer-illustrator*

Us	Cats
Primarily verbal and visual communicators	Rely heavily on scent and subtle body language for communication
"Intellectualizers"—highly cognitive, processing information internally for sometimes lengthy periods before acting on it	"Reactors"—react quickly and instinctually to stimuli and information in their environments
Often focused on the future or the past	Very present-oriented
View events as stories—ongoing narratives with beginning, middle, and end	View events as individual stimuli, to be reacted to immediately
Hold grudges, plot revenge	Remember but don't hold grudges
Can adapt to a "command and control" social structure	Hah!

The Secret Cat Code

Throughout this book, you'll discover numerous ways to persuade your cat that what *you* want is also what *she* wants. Here's your SECRET CAT CODE for outsmarting cats:

- *S*afety

- *E*nvironmental enrichment

- *C*onstant communication

- *R*ewards, routines, rituals

- *E*nough of everything

- *T*rust

- *C*onsistency

- *A*cceptance

- *T*olerance

- *C*almness

- *O*bservation

- *D*iversion

- *E*xclusion

What Doesn't Work?

There's very little that always works with cats. What *never* works?

- Punishment

- Force

- Violence

- Anger, rage

- Belittling, insults

- Revenge

- Commands

WHAT'S WRONG WITH PUNISHMENT?

Punishment just doesn't work with cats. Pushing, swatting, or hitting a cat will tend to make her fear and dislike you, and to avoid your company in the future.

Punishment can form unwanted and unintended associations in your cat's mind. Even in the unlikely event that you catch her within seconds of scratching your stereo speakers, punishment may do more harm than good—especially if the "misbehavior" is a normal cat behavior (like scratching). She'll have no idea why you're upset and may decide you're untrustworthy or just nuts—someone to avoid.

Your cat's "window of correction" is *3.5* seconds, *at most.* That's the amount of time you have between your cat's action and any chance of convincing her that your appalled reaction is a response to it. You have to catch her in the act. But even that's no guarantee, because . . .

Your cat may associate your sudden weird or angry reaction with something completely different—the presence of another cat or person in the room, or the current phase of the moon.

Even if your cat understands that you don't approve of something, punishment won't make her stop. If she finds the activity pleasurable or necessary (like scratching), she'll continue to do it—but only when you're not around.

Punishment may "reward" your cat's actions. She might continue the behavior to get more attention from you (or just to watch you fly off the handle).

WHAT ABOUT AVERSIVES?

Many cat books and experts advise using "safe aversive training techniques." These "aversives" are devices, techniques, and strategies that try to modify your cat's behavior by training her to associate a specific behavior with something negative, unpleasant, or even mildly painful. Aversives commonly recommended include:

- Squirt guns: squirting the cat when you catch her doing something you don't like (like scratching the sofa).

- Sharply tapping the cat's nose. (In theory, this imitates MomCat's training technique in which she raps a misbehaving kitten's nose with her paw.)

- Holding the cat by the scruff of the neck and shaking her gently (thought to imitate MomCat).

- Electric shock mats placed on furniture, on kitchen counters, or in doorways. These mats deliver a harmless but unpleasant electrical shock when the cat steps on them.

- Alarms: motion-detecting devices that sound a piercing or unpleasant sound, or deliver a spray of water or an unpleasant-smelling substance when triggered by the approach of a cat.

- Homemade noisemakers, such as soda cans filled with pennies or pebbles and taped shut. If stacked precariously on tabletops, kitchen counters, or other locations, these tumble noisily if the cat jumps and disturbs the stack. In theory, this unpleasant surprise teaches her to avoid that location.

- Owner-operated noisemakers, such as air horns or rattles.

These *can* work, at least in the short run. In *Outsmarting Cats,* you'll see similar techniques listed along with other options for modifying behavior. But in general, I don't recommend aversive or negative techniques for modifying your cat's behavior. I just don't like them.

Why? It's simple, really. When it comes to your cat, "nice" matters.

"NICE" MATTERS

Your cat is an intelligent, adaptable, and polite creature. Ambushing her with unpredictable, unpleasant booby traps is rude and may well be counterproductive. The best way to ensure that your cat will be the best-behaved and happiest animal she can be is to do everything you can to ensure that she can trust her home—her territory—to be a predictable, comfortable, safe, low-stress environment. A cat subjected to unpredictable electric shocks, piercing alarms, noisy cascades of falling objects, jets of water, or other nasty surprises is at risk of becoming a fearful, wary, suspicious, and stressed cat. Such negative experiences will gradually erode her sense of confidence, safety, consistency, and comfort in her environment.

In the unforgiving world of the wild, a small cat needs to be tough, solitary, wary, and suspicious. To survive, she suppresses the softer, gentler aspects of her nature. When we bring cats indoors, we want to replicate, as far as possible, the positive aspects of the free, unfettered life—opportunities for mock hunting, vigorous exercise, climbing, foraging, and adventure—but not its negative, unpleasant, dangerous, stressful aspects.

Don't booby-trap your cat's world. Offer her instead a safe, cat-centered environment of peace and plenty. Her home should be a haven where she feels confident and secure enough to develop and display the sweeter, more complex aspects of her personality, and where she can bloom into the best cat she can be.

Outsmarting Cats offers plenty of alternatives to aversive and negative training techniques. Try positive reinforcement, distraction and diversion, redirection, exclusion, and the many other proven cat-outsmarting strategies first. If you do choose to try aversive training techniques, use them for as short a time as possible.

What Does Work?

PREVENTION

Undesirable cat behavior is always easier to prevent than fix. Once medical causes are ruled out, it's important to find and correct the root causes of your cat's unhappiness and her unwanted behavior, before it gets out of hand and becomes a habit. Bad behavior can become ingrained as quickly as good behavior, and bad habits can linger long after the original problem is corrected. The quicker you discover and solve the problem, the better.

DIVERSION

Your cat is forever on the lookout for a better deal. So when she misbehaves, give her one. Fight brewing? Instead of yelling or throwing your shoe, grab an interactive toy and start a play session. Cat scratching the sofa? Toss a Ping-Pong ball or treat past her, leading her away from temptation. She'll likely give chase and forget about the sofa.

Don't try to divert her by paying her special attention, though—this might reward and reinforce her bad behavior. The diversion should be something neutral (not associated with you) but attractive to your cat. A small, handheld laser pointer is an ideal diversion tool. Most cats adore chasing that little red dot. Keep one handy in your pocket.

VERBAL CUEING

Cats are exquisitely sensitive to tone of voice and variations in inflection. Verbal cueing works if used consistently. Yelling in anger or frustration will only confuse your cat, or make her fear and avoid you. Devise a particular way of saying your cat's name, and use it only when she misbehaves. For example, use an exaggerated upward inflection.

Or invent a cue phrase and method of delivery, such as a sharp "No, no, no, cat, cease and desist!" Reserve this cue phrase *only* for a particular misbehavior, and use the phrase consistently and immediately whenever you see that behavior. How well this works depends on the cat, how highly motivated the misbehavior is, and how consistent you are. If you have multiple cats, make sure their names (or nicknames) sound sufficiently different.

Cat Names

"They say the test of literary power is whether a man can write an inscription. I say: Can he name a kitten?"
—*Samuel Butler*

An astonishing number of cats are known simply as "Cat," "Kitty," or "Baby." Of course, what a cat calls herself—what poet T. S. Eliot called her "Deep and inscrutable singular Name"—remains a delicious, eternal mystery.

According to one survey, an increasing number of cat owners give their cats people names, reflecting their status as full family members. Top names include Max, Molly, Maggie, Daisy, Lucy, Bailey, Sadie, Ginger, Charlie, Chloe, Sophie, and Sam.

Another survey found that the most popular names for male cats are Smokey, Tiger, Max, Charlie, Rocky, Sam, Sammy, Mickey, and Toby; the most popular names for female cats are Samantha, Misty, Muffin, Fluffy, Patches, Pumpkin, Missy, Tabitha, and Tigger.

C'mon, cat owners! Let's get a little more creative!

Pumpernickel? Pochade? Pernickety? Plum Goddess? Razzmatazz? Baby Jaguar? Glossie Flossie? Chrysanthemum? Sir Sterling Silver Longfellow? Pemigewasset Q. Feathers?

The king of feline nomenclature, claiming the longest known cat name, was "The Most Noble the Archduke Rumpelstizchen, Marquis Macbum, Earle Tomemange, Baron Raticide, Waowler and Skaratchi," known as "Rumpel" to his friends. Rumpel lived with the nineteenth-century English author and poet Robert Southey.

REMOVING THE AUDIENCE

When your cat is doing something you don't like, quietly leave the room. Don't say anything at all. Often, in removing the audience, you remove the motivation for the behavior. Leaving also helps to interrupt the unwanted behavior pattern without reinforcing. Wait several minutes. Then, in another location, initiate a game or play session.

Similarly, if your cat is on your lap, kneading your leg painfully with her claws, get up quietly and let her drop gently to the floor. Leave the room for a few minutes. This way, you're not telling her that you don't appreciate this affectionate gesture (reminiscent of the mother–kitten relationship), but are simply interrupting a pattern you don't want to reinforce. Whatever you do, *don't* yelp or squeal in pain. Your cat may think your kittenish vocalization is part of a game.

GIVING IN

Sometimes the best way to outsmart a bothersome cat is to give in. Cats are often satisfied with brief, if intense, interactions. When your cat hops up on your lap while you're reading or typing on the computer, don't exclaim in annoyance and push her away. Instead, give in. Indulge her for a few minutes. Devote your full attention to this magical, beautiful creature who has sought you out. In most cases, a few minutes of petting and praise is all she needs, and she'll saunter away, happy. You'll likely find the brief break has done you good, too.

Giving in also works when your cat insists on sharing your dinner. The more you try to shoo her away and keep her off the table, the more she'll insist. The secret to outsmarting her? Give in. Let her at your dinner plate. Chances are, she'll be satisfied with a brief sniff or a quick lick. Cats are seldom interested in actually eating more than a token amount of people food—they're just extremely curious about it because *you're* eating it! (Hint: If you're worried that she'll really run off with your dinner, ensure that she takes no more than a brief lick by sprinkling a bit of pepper on it.) See chapter 8 for more tips on outsmarting food-snatching felines.

EXCLUSION

Sometimes the best way to outsmart a problem behavior is to remove from your cat all temptation. Close off the room where the misbehavior occurs. Relocate breakables and valuable items to a cat-free zone. Exclusion is a valuable safety strategy, too. If you're working with toxic materials, polishing the silver, repairing window screens, using power tools, scrubbing the bathtub—shut the door and keep your cat out until all is back to normal.

Boredom and Misbehavior

Insistent paw taps, escalating to claw rakes, at midnight. Howling at 3:00 a.m. Toilet paper unrolled and dragged everywhere. Trash tipped over. Papers chewed, pencils knocked off desks. Screens clawed. Stealth attacks on ankles. Has your cat gone bonkers? Is it revenge? No. She's bored.

Boredom results in a whole range of annoying, puzzling, and maddening feline misbehavior. Paradoxically, cats love routines, but they also crave novelty and mental, physical, and emotional stimulation. If there's too little excitement and adventure in their lives, they'll stir up some of their own. You might not approve of their choices.

Quite often, feline destructiveness—chewing; relentless bothering of people trying to read, cook, or sleep; crying and racing about at night; and similar mischief—are caused by boredom. Zoo animals who don't have enough interaction with fellow creatures suffer from what's known as confinement stress, leading to depression, "stereotypies" (repetitive pacing, self-mutilation), and other serious behavior and health problems. If zookeepers don't provide regular social and environmental enrichment, these animals risk leading an empty, biologically purposeless existence. A cat left alone too much is similarly at risk for boredom, loneliness, and stress.

THE TRADE-OFF

A cat out on her own is a busy, active, constantly stimulated animal. She must define, patrol, and defend her territory. She must be on guard against threats, from wild weather to predators to competition for resources. If she doesn't have access to human-provided food, she must locate, stalk, hunt, and kill prey. Even asleep, a cat walking "by her wild lone" keeps an ear cocked for danger or opportunity. Unspayed female cats bear litter after litter of kittens whom they must guard, feed, nurture, and train.

These are natural activities for which cats are superbly designed and evolved. But the perils are considerable. In taking cats into our homes, we make a trade-off. They give up their natural—if perilous and likely short—life, for a much more

Does Your Cat Have a Secret Life?

Jürgen Perthold was curious about where his free-roaming cat, Mr. Lee, went and what he did all day. But the automotive engineer and inventor did more than wonder. He built the world's first "CatCam." He equipped a tiny digital "keychain" camera with a microcontroller, wrote software to control the camera, and strapped the device onto his cat's collar.

The shockproof, scratchproof, watertight, theft-resistant camera takes one picture per minute, for forty-eight hours. The whole setup weighs just 2½ ounces and didn't bother Mr. Lee at all once he got used to it. For feline safety, Perthold advises the use of a breakaway collar, even though the camera could then be lost.

The CatCam is just one example of the brave new world of "furveillance" technology. Miniature, inexpensive, highly reliable, and extremely capable cameras can record your cat's activities either in internal storage (such as a removable SD card) or send a live feed via wireless connection direct to your computer, PDA, cell phone, or other device. The technology—tiny video cameras and GPS location recorders—is the same as that used for tracking packages and vehicles.

"Furveillance" at Home

Do you wonder which of your four cats is using the floor instead of the litter box? Does your recuperating cat need a watchful eye? Want to look in on your cat while you're at work? Why is your cat gaining weight even though she's on a diet? (Maybe because she's sneaking out the door for extra meals at the neighbor's house or doing a little freelance hunting.)

You can now set up a room-panning video camera in your home that records activities in that room, while transmitting a live feed to a remote location. Log on at work, and you can watch your cat snooze all day. One camera is never enough; many pet owners equip each room or area with its own camera so they can follow the activity throughout the house. There are garden-cams, basement-cams, barn-cams, and even food-bowl cams.

Petcams are now a must in many pet-related businesses, such as animal hospitals and boarding facilities, because clients demand them. Some veterinary clinics have cameras so clients can watch pet surgeries and keep an eye on their recuperating pets. There is a growing array of choices, both wired and wireless, plus night-vision cameras (to track the action in your yard) and cameras equipped with two-way audio. (See Resources.)

With the new technology, your cats need never be out of your sight, and they can bring back a record of their independent adventures. What new secrets will we learn about these most familiar creatures?

comfortable, healthier, and probably much longer life. By making this trade-off on their behalf, though, we also assume a big responsibility. We need to provide safe, acceptable, cat-friendly substitutes for all those natural activities in our own homes and sources of daily stimulation that are so essential in maintaining feline mental, physical, and emotional well-being.

Happily, you're about to discover that this is not mission impossible. *Outsmarting Cats* will show you that it's doable—and it's fun. Creatively optimizing your home for your cat's needs will enhance your cat's health and happiness, strengthen your mutual bond, and help make the adventure of cat ownership a pleasure and a joy.

OUTSMARTING BOREDOM

Cats are active, agile, athletic, high-energy, highly tuned creatures—the high-performance race cars of the animal world. Don't make your cat lead the life of a potted plant.

- One of the best cures for feline boredom is a companion cat. Ideally, adopt a pair of compatible cats or kittens.

- Cats crave the excitement and stimulation of hunting—or regular mock-hunting play. Offer plenty of interactive playtime. Use a fishing-pole-type toy, put away between play sessions to keep the fun fresh.

- Spend quality cuddling-and-conversation time with your cat, every day.

- Place an inviting cat tree, cat hammock, or perch near a window (perhaps on a wide windowsill) and set up a bird feeder outside. (Avoid live caged birds. Even if your cat can't get at them, her presence will likely stress them.)

Tabigail is an avid bird-watcher.

The Best Cat Toys*

1. *The Feline Fisher (successor to the Galkie Kitty Tease):* Voted Best Cat Toy Ever by the Bobcat Mountain Cat Toy Test Team. This interactive fishing-pole-type toy is deceptively simple, but cats go crazy for it. It's also incredibly sturdy. Our original, fifteen-year-old Kitty Tease is still in heavy daily use. The Feline Fisher looks like it will last even longer.

2. *Da Bird:* The best of the "flying-bird" fishing-pole toys. It makes an amazingly realistic bird-flying sound. Be sure to get replacement birdies.

3. *Fly Toy* (fishing-pole toy): Made with genuine fishing flies (no hook, of course), these skitter around tantalizingly at the ends of their strings like insects. There are several different models. All are big favorites on Bobcat Mountain.

4. *Whirly-Bird Cat Exercise Toy:* This interactive flying-bird toy has a two-piece bamboo pole. The feather bird has a foam "body" that cats *love* chomping down on when they make a kill.

5. *Kitty Can't Cope Sack:* This sturdy, well-made small cotton sack filled with 100 percent fresh catnip has remarkable staying power. It's great for bored and stressed-out kitties. They chew it, rub it, bite it, stalk it, carry it around the house, and sleep atop it. And it just keeps going and going and going . . .

6. *Play-N-Squeak* (microchip sound toy mouse): This is a soft, furry mouse with a sound-producing microchip inside. When the cat bats it or pushes it, it makes an eerily mouse-like squeak. This can be hung from a cord on a door, or used as a solo floor toy without the cord.

7. *Enchantacat catnip-filled toys:* These are some of the best-made and sturdiest catnip toys we've found. Some are washable and refillable. The catnip they're filled with (which can be purchased separately) is really primo stuff, according to the connoisseurs at Bobcat Mountain.

8. *Alpine Scratcher:* This corrugated cardboard scratch pad comes in an angled holder, like a mini ski slope. A supply of Cosmic Catnip is included for rubbing and sprinkling on the scratch pad. Inside the holder is a dangly toy and a space that's perfect for hiding treats and toys for cats to discover on their own.

9. *Hot Cats:* These catnip-filled "cigars" are just the right size to chew, bat, or just sit on.

10. *Pawbreakers:* These balls of compressed catnip are about the same size as a Ping-Pong ball and make a satisfying rumbly sound as they roll across a hard floor. The test team never seems to get tired of batting these things around!

*As determined by the Bobcat Mountain Cat Toy Test Team.

Catnip

From ancient times, catnip has been used as a human medicine, tea, and tonic. But for cats, it's more of a recreational drug. Their ecstatic rolling and rubbing is a response to cis-trans-nepetalactone, the major component of catnip's volatile oil. In cats, the odor of catnip affects the same biochemical pathways as marijuana and LSD. Chemically, the catnip trip appears to be a true psychedelic experience, lasting anywhere from five to fifteen minutes.

Catnip is a safe, self-limiting drug. After a catnip "trip," even a strong responder won't react to it for at least an hour. Overexposure can lead to lessened sensitivity or indifference.

Sixty to 70 percent of adult cats respond to catnip to some degree; sensitivity is inherited. Male cats tend be more strongly affected than females. Kittens under six months avoid it. Catnip's appeal is in its scent rather than in its taste. A sensitive cat, using her vomeronasal organ (her "taste-smell" sense), can detect the volatile oil of catnip at an atmospheric concentration of one part per billion.

Using Catnip to Outsmart Your Cat

Catnip can be a powerful tool in behavior modification. Use it to reward your cat after a grooming session, medication, or a veterinarian visit. An occasional catnip party in your cat's open carrier will help make the carrier a place of fun rather than stress. Some veterinarians give catnip to cats on the examining table to distract them and help remember the visit as a pleasant experience.

Dried catnip rubbed on a scratching post, or placed on a platform on a climbing tree, can make it more attractive, diverting your cat from alternatives.

- Provide a securely lidded tank of colorful tropical fish.

- Set up several secret hideaways, furnished with comfy pillows or old sweatshirts that carry your scent.

- Some cats are fascinated with "cat videos" featuring scampering squirrels and fluttery birds; others aren't impressed.

- Offer her inner wildcat foraging practice. Scatter and hide some treats (plain dry cat kibble, perhaps a different brand, shape, or flavor than her usual chow) for your cat to discover on her own.

- Use a "treat ball," a special toy that holds kibble or treats. Your cat has to roll around or otherwise manipulate the toy to get at the food.

- Place her toys in a toy box. Rotate toys frequently, so they always seem fresh and new. Avoid leaving pre-filled catnip toys out continuously; too much

Small catnip-filled toys can serve as handy diversions from unacceptable activities. Keep one handy, ready to deploy as needed. A catnip party is a great stress reliever and antidote to boredom.

Grow Your Own

Nepeta cataria is a hardy perennial, a sun-loving member of the mint family. Sturdy plants grow two to four feet tall. Grayish green, heart- or spade-shaped saw-toothed leaves, covered with fine, whitish down, are set opposite on the square stem characteristic of mints. Vertical spikes of purple-pink flowers appear atop the plants in July and August.

Catnip is easy to grow, either indoors in pots or in the garden. Outdoors, plant in a sunny area. (Plants grown in shade produce less volatile oil.) Catnip can be invasive, spreading rapidly by both seeds and underground runners. For maximum oil content, pick leaves in the morning just after the dew has dried. In season, offer your cat fresh leaves.

Indoors, catnip is easy to start from seed. Place pots in a sunny window in a cat-free zone. If your cat has access to the whole plant, it won't last long.

To dry catnip, cut leaves before they've yellowed. Spread them out on brown paper in a dry, shady spot; or, hang whole branches upside down. When fully dry, store leaves in a cat-proof container in a dark, dry location. *Never underestimate your cat's cleverness, tenacity, and zeal in trying to gain access to this prize.* Plastic bags just won't do. I use gallon-size, unbreakable plastic food-storage jars with screw-top lids.

regular exposure to the herb desensitizes a cat to its delightful effect—a real shame.

- About once a week, throw a catnip party. Fill an old sock with fresh or dried catnip leaves, or just sprinkle the leaves on the floor (you can vacuum them up later) and let the good times roll.

- Blow bubbles. A simple, inexpensive, widely available child's toy can let your cat enjoy pursuing a new kind of prey, and help you feel like a kid again. Even if you haven't touched a bubble wand since fourth grade, you'll rapidly recover the knack. Your cat will adore chasing the sparkly bubbles, pouncing on them, and then searching for the vanished prey. Bubbles can be messy, leaving sticky soap spots on the floor. Enjoy your bubbly fun in the kitchen or other linoleum or tiled floor area, and damp-mop afterward.

Away at work all day?

- Establish a morning wake-up ritual and an evening homecoming greeting-and-play session.

- Leave a radio tuned to a station with a mix of soothing voices and music (a classical music station is ideal).

- Leave a tape with a loving message, in your own voice, timed to play during the day.

- Treat your bored cat to a cat-friendly video. A variety of videos are marketed specifically for cats, but those aren't the only ones she might enjoy. Try a tape of a cozy, crackling fireplace (safer and cleaner than the real thing) or an aquarium video featuring tropical fish and waving plants. Offer nature and wildlife films, especially ones that spotlight fluttery birds, scampering squirrels, or your cat's wild relatives. If you have to be away for several hours, set up a video recording on a timer to play it later, or play it as a continuous loop.

Kitty Time-out: The Safe Hideaway

One of the nicest, and smartest, things you can do for your cat is to give her a room of her own. Just like us, cats sometimes get stressed out by the noise and bustle of life and need a place they can chill out. You cat's "safe hideaway" should be a haven she seeks out on her own and enjoys spending time in. When might the safe hideaway come in handy?

- Parties. Even sociable cats get overstimulated.

- Holidays. Decorations, unfamiliar scents, and visitors can be too much.

- Fireworks and other noisy celebrations.

- Thunderstorms. Some cats are extra sensitive to these.

- Houseguests. They can wear on your cat as much as on you. If your cat isn't accustomed to children, their loud, high-pitched voices can be unnerving.

- Remodeling, repairs, when the plumber comes to call.

- Recovering from illness. Peace and quiet speed recovery after surgery, injury, or illness.

- Comfortable confinement while you diagnose (and clean up) litter box misbehavior. *This is prevention, not punishment.*

- Retraining. When your cat has been misbehaving, a change of scenery and reduced temptation can lower both your and her stress levels.

- Anytime your cat's feeling stressed or just needs a break.

Your cat's safe hideaway can be a small bathroom, a spare bedroom or guest room, laundry or utility room, or basement. An ideal spot is where one of your litter box stations is set up. In a pinch, a large pet carrier, cage, or dog crate will do. It should be ready for service at a moment's notice. Add a bowl of fresh water and a small bowl of dry food. If your cat will be in residence more than a very short time, add a comfy bed or blanket, a few toys, and a small portable scratching post. Check up on her, but be sure she doesn't escape until everything's back to normal.

Your cat may also find one or more safe hideaways—maybe a closet, or a cozy corner somewhere—on her own. Respect her preferences. Avoid bothering her when she retreats there, unless it's really an emergency. And wherever she chooses to retreat, make sure she has easy access to a litter box and other kitty necessities.

To help mask loud or unfamiliar noises (revelry, hammering, power tools, yelling kids), play a radio, CD player, or MP3 player. Talk shows and classical, harp, and new age music are all great "white noise" for masking loud, harsh, disturbing sounds.

HIDEAWAY SAFETY

Before introducing your cat to her safe hideaway, check carefully for dangers: toxic or poisonous plants, breakables, cleaning or other household products, loose electrical cords. Install child-proof latches on all cabinets, cover unused outlets with covers, and keep anything even remotely dangerous stowed. Make sure all window screens are intact and secure.

It's okay to be fanatic about your cat's safety. Advise guests, visitors, or tradespeople that your cat's in residence in her hideaway and is not to be disturbed. Make a big deal about it so they'll remember. Post a prominent sign at eye level on the door of the hideaway: DO NOT OPEN THIS DOOR—CAT IN RESIDENCE. If you really want to make an impression, have it read: DO NOT OPEN THIS DOOR EXCEPT AT YOUR OWN RISK! VICIOUS, MAN-EATING PREDATOR IN RESIDENCE! For extra protection, hang an additional sign on the doorknob.

When you're ready to let your cat out, do an outside-the-hideaway safety check first. Make sure all party food, construction materials, and other hazards have been picked up and put away. Check doors and windows.

For your cat's pleasure, convenience, and psychological health, it's a good idea to leave her hideaway accessible to her at all times, for whenever she feels the need to retreat from the household hustle and bustle. Spending time in her hideaway should be a happy, peaceable experience she likes well enough to seek out for herself when she needs a break—never a punishment. When you take her to her hideaway, lay on the kisses and lavish praise. The safe hideaway is not just for physical safety. It's also a great everyday stress reducer and a safety valve for overstimulation.

A Plea for Connection

> "By associating with the cat one only risks becoming richer."
>
> —*Colette*

Veterinary ethologist Myrna Milani, DVM, marvels at how well our cats adapt to living and thriving in human homes. That they do so is a tribute to feline intelligence, perceptiveness, and emotional flexibility, plus their ever-ready nose for a good deal. But there's something else.

Cats seem to *like* us!

Perhaps we amuse them. Perhaps, in their mysterious feline way, cats trade anecdotes about silly human tricks and "the dumbest things my people ever did." What must they think of the odd habits and preoccupations of these vertical, furless, tail-less (but handily can-opening) beings, so easy to figure out and so easy to manipulate?

Your cat's favorite toy, and best guarantee against boredom, is you. Boredom-caused destructiveness or misbehavior is often a plea: *Pay more attention to me!* It's a loving plea, called across the interspecies barrier, for play, cuddling, conversation, companionship, interactivity—for closeness and connection.

Angel relaxes in a comfortable, low-stress environment.

Quick Tips

1. *Reduce stress in your home.*

   - Speak softly. Avoid yelling.

   - Walk slowly and softly.

   - Stay organized.

   - Keep noise down.

   - Take off shoes in the house.

2. *Plan ahead.* While life brings inevitable changes, include your cat's needs and feelings in your plans as much as possible.

3. *Don't set bad precedents.* If you won't like a behavior in full-grown cat, don't encourage or reward it in a kitten. Examples are rough play with hands and toes, and "ankle wrapping." *Never give your cat any reason to think of your hands (or any human's hands) as either toys or weapons.*

4. *Minimize changes in routine.* In times of change, maintain as normal a schedule as possible, with regular mealtimes, litter box care, and playtime.

5. *Introduce all changes slowly and gradually.* Cats are extremely sensitive to change because in the wild, change means threat.

6. *Introduce new people and animals gradually.* Never just throw your cat together with a new person or animal, or (worse) shove her in a newcomer's face. This can overwhelm a cat—it's just too much, and claws might swiftly be deployed. Never force her to interact with a new person or animal before she's ready. If the newcomer is expected, bring your cat an object carrying that person's or animal's scent, so she can gradually become comfortable with it.

7. *When introducing two cats, your wisest course is to give them plenty of time and space to make their own social arrangements.* Observe discreetly, though. And when any new person or animal moves in, keep alert even if your cat doesn't react immediately. It might be a while before the stress or stimulus level goes up enough to unhinge her.

Cats are great at sharing—as long as it's *their* idea.

8. *Be scent-sitive.* Pay particular attention to scent in your cat's environment: your personal scent (including cosmetics and personal care products); other animals, including cats outdoors; and visitors and the scents they bring in on their clothes and shoes. Cats recognize individuals by scent, not visually.

9. *Allow cats to share as they wish.* Cats are adept at sharing, as long as they have a sense of abundance and continuity. A favorite chair might be claimed by one cat in the morning and a different cat after dinner.

10. *Don't force sharing.* Provide individual bowls and multiple feeding stations. If you have small children or dogs, feed your cat on a countertop, or on the top of the refrigerator. The number, location, and cleanliness of litter boxes is critical in both preventing intercat prickliness and settling down any conflicts that do erupt. Always provide at least one litter box per cat, plus extras, in multiple locations if possible.

11. *Listen.* Learn your cat's vocabulary: her many meows, calls, and purrs. Learn to distinguish between growling and purring; the differences can be subtle.

12. *Be a good provider.* When your cat seems nervous, aggressive, or unhappy, or is misbehaving, look at the situation through her eyes. Which important resources might she feel are in jeopardy? Reassure her by providing plenty of everything, in an atmosphere of peace and calm.

13. *Watch for signs of trouble brewing.* Don't ignore litter box misses, fighting, hiding or sulking, or repeated sparring between cats. Watch body language for staring, hissing, flattened ears, pouffed-out fur, "Halloween cats," or lashing tails. The longer you ignore misbehavior or let it persist, the harder it is to correct.

14. *Be ready to take action.* At the first signs of trouble (marking, spraying, growling, cats staring at each other), set up at least one extra litter box station. Often this will settle them down.

15. *But don't meddle.* Unless a clearly dangerous conflict is brewing, or one cat is being constantly harassed, take a benign, hands-off approach. Feline social structures are dynamic, ever changing. Some cats, for reasons of their own, choose to stay outside the main group. Let them.

16. *Let your cats work out minor differences among themselves.* Although it's tempting, avoid interfering in brief intercat tiffs. They may sound

alarming, with lots of hissing, growling, and squealing. The fur may fly for a moment. But most of these "fights" are minor and transient. Meddling may simply add fuel to the fire and make things worse.

17. *Don't obstruct normal feline behavior.* No matter how peculiar you find such feline habits as sniffing one another's (or your) body parts, washing sensitive areas, licking, and so on, don't interfere or loudly express disgust. These are important communication and bonding rituals.

18. *Avoid unpleasant surprises.* Never sneak up on a cat, startle her, or grab her unaware. Especially, don't interfere with a cat who's sleeping, washing, eating, or using her litter box.

19. *Keep cool.* Your cat is stimulus-driven. Interacting with her while you're angry, stressed, or tense is like pouring gasoline on a fire. She'll instantly pick up on the intensity and negativity of your feelings, making things worse. So when she annoys you for the zillionth time in one afternoon, take a deep breath and calm yourself first. Remember: Calm owner, calm cat.

20. *Be consistent!*

"Cats do what they do because it works!"

—*Myrna Milani, DVM*

Chapter 5

The Litter Box Blues

"A man has to work so hard so that something of his personality stays alive. A tomcat has it so easy, he has only to spray and his presence is there for years on rainy days."

—*Albert Einstein*

A Frustrating and Deadly Problem

In 1995, the National Council on Pet Population Study and Policy set out to learn why people surrender pets to animal shelters. They discovered that in over one-third of all cases, it was behavior problems that shattered the human–pet bond.

The NCPPSP compiled a list of reasons cat owners give when surrendering adult cats. The seventh-most cited reason was "house soiling." Experienced shelter workers know, though, that reasons #1 ("too many cats"), #2 ("allergies"), and #3 ("we're moving") are often proffered to conceal surrendering owners' embarrassment, frustration, and rage over the real reason—their home has turned into a giant, stinking litter box.

Anyone whose cat has ever decided to "think outside the box," however briefly, can sympathize with those deeply frustrated cat owners who feel their only solution is to give away their cats. There's no cat behavior problem as frustrating, maddening, and challenging as what exasperated owners describe as "peeing all over the house."

Why, Cat, Why?

Your cat really doesn't want to urinate on the carpet or defecate in the closet. But sometimes, for some reason, a cat feels he has no choice. Perhaps a rival is guarding the only available litter box. Perhaps the box is filthy. Perhaps he's in pain. Perhaps he's ill, and can't get to the box in time. You *can* be sure that your cat isn't misbehaving out of sheer spite or nastiness, or to get revenge on you because he hates your new boyfriend.

He's doing what he's doing because it *works*—it solves his problem. He may not be any happier with the solution he's found than you are. Cats are clean, fastidious animals. In the wild, they take great pains to conceal evidence of their presence and activities. Your cat may be working awfully hard to find spots within your home that are sufficiently private and hidden, and he may be experiencing tremendous frustration and insecurity about his situation. If his problem is caused by illness, he's likely in pain as well.

Prevention: Be a Good Provider

By discovering his dilemma as soon as possible, and offering acceptable alternatives, you aren't just outsmarting him and solving his (and your) immediate problem. You're doing him the great favor of restoring his comfort and confidence in his environment. By helping him *gently* through this crisis and seeing the situation from his perspective, rather than overreacting with rage and frustration, you'll solve the problem quicker, together.

Your best efforts at prevention may not always be completely effective. Still, one of the smartest things you can do to prevent litter box problems is to be a good provider:

1. *Provide plenty of boxes, in multiple locations.* A good rule is "One box per cat, plus one extra." Set up the boxes in at least two different sites, or "stations." Perhaps you could place a few boxes in a quiet spot in the basement, and another few in an upstairs bathroom or utility room. Even if you have only one cat, you should provide at least two boxes, preferably at two different sites. This is especially true if your home is large or has multiple floors, or if your cat is very young or elderly.

2. *Select litter boxes with your cat's needs in mind.* Most pet store boxes are too small. Cats need plenty of room to stand up, turn around, dig vigorously, and bury their waste completely. Instead of pet store boxes, get large (at least sixteen by twenty-two inches and twelve inches high; larger is better) clear plastic storage boxes, sold in discount stores and home centers for storing clothing. These are generally less expensive than pet store boxes, and most cats appreciate the extra size. Look for

boxes with relatively smooth, flat sides and bottoms. For young kittens and elderly cats, provide some low-sided boxes, too.

Avoid covered boxes or "stealth" litter boxes disguised as potted plants, cabinets, or benches. Among the problems with covered and disguised litter boxes:

- They're too small for many adult cats. Cats can't stand up or turn around, and have no place to put their tails.

- They trap odors inside (distasteful for cats).

- They're not as escapable as open boxes. A cat may avoid such a box because he fears being trapped or ambushed by a rival or bully.

- "Out of sight, out of mind." If you don't see (and smell) the box and whether it's been used, you're more likely to neglect regular scooping.

- The (sometimes considerable) expense of these contraptions may make you less inclined to provide as many boxes as you should.

Be wary of elaborate, high-tech, "self-cleaning" boxes and "litter systems." This is not rocket science. All those latches, snaps, grilles, ledges, levers, moving parts, and other "improvements" trap wet, smelly litter and make the box harder to take apart and clean regularly. Remember Mewphy's Law: "Self-cleaning litter boxes aren't." Simpler is better.

3. *Use unscented clumping litter.* Cats are *extremely* sensitive to the texture of their litter. Cats favor a soft, sand-like texture, which most closely resembles what they'd use in nature. It's also the most economical and easiest to keep clean. Avoid pebbles, pellets, and anything that would feel sharp and uncomfortable on your cat's paws.

   Avoid scented litters and those with "odor control" ingredients— usually overpowering flowery scents that'll likely have *you* wrinkling your nose in disgust. A litter box shouldn't smell "like" anything— except your cat. It should have a neutral smell of clean sand or clay, complemented by your cat's familiar "this is mine" odor, perceptible to him if not to you.

   If your cat straddles the litter box, perches up on the edge, or shakes his paws excessively after using it, he probably doesn't approve of your litter choice. Take this as a warning sign.

Flush with Success?

You've probably met at least one cat owner bragging about their toilet-trained cat. Some cats have been trained to use toilet facilities designed for humans, and a few have even been known to imitate their owners and figure out the process for themselves. But before you go adding up your savings in cat litter, consider your cat's temperament, wishes, and preferences.

There are kits and videos available to help train your cat to use a toilet instead of a litter box. If you feel your cat might enjoy the challenge, or if you would, go ahead. But remember: Wise owners never insist.

Hoping to grab a piece of the billion-dollar litter market, manufacturers are forever introducing new products, materials, shapes, and textures. These are generally aimed at owners, not cats. You may personally prefer natural or organic products. If your cat agrees with you, fine. But the best litter for your cat is one he'll use consistently.

4. *Use enough litter.* Keep about four inches of litter in each box. This gives your cat plenty of litter for digging and burying, and makes your scooping job easier. That's because the clumps tend to "float" on the litter rather than sticking to the bottom of the box, where scooping them is messier and more difficult.

5. *Be consistent in your choice of litter.* Once you find a brand of litter your cat likes, stick with it. Filling the box with "whatever's on sale this week" is a dangerous form of false economy. Your cat is a creature of habit, especially when it comes to the litter box. That familiar odor, texture, and paw-feel are extremely important.

6. *Keep the litter clean.* Scoop all boxes at least once a day. (See "Do's and Don'ts for Litter Box Care," page 108.) Easy access and visibility make maintenance simpler and quicker.

7. *Replace boxes periodically.* A well-used litter box accumulates scratches that trap odors and bacteria, making it harder to scoop waste neatly. When a litter box is starting to look scratched up, throw it out and replace it with a new one. (This is another good reason to avoid expensive boxes and "litter systems.")

8. *Don't use "anti-tracking" mats,* which are plastic or rubber mats with spikes or other protrusions designed to pull bits of litter off a cat's

paws. Most cats hate these because they're uncomfortable or painful to walk on. The presence of one will cause many cats to avoid the litter box area entirely and take their business elsewhere.

LITTER CARE SUPPLIES

The simpler your litter box setup, and the easier it is to find everything you need when you need it, the more often you'll be inclined to do the job. This will mean: a cleaner, tidier, better-smelling home; less time wasted looking around for your tools; and most important, a happier, better-behaved cat!

Keep everything you need for daily maintenance at each litter box station:

- A sturdy litter scoop. Get a large, sturdy model made of stainless steel or tough plastic. (Gadget fans will love the battery-powered scoop that does the shaking for you.)

- A paint scraper. Get a three- to four-inch metal-bladed one from your local home center or hardware store.

- A supply of clean litter—unscented clumping litter is best.

- A small covered trash can (the sturdy plastic tubs some cat litter is sold in work great, too) lined with a couple of inexpensive plastic bags. You can use plastic supermarket bags, but use two or, better, three layers. They're not designed for such tough duty. Wet, used litter is heavy! This "lined trash can" setup is great when you're cleaning up for several cats.

- Instead of a trash can, you can keep a supply of gallon-size, ziplock plastic freezer bags on hand. Scoop the waste into the bag. After each cleanup session, just zip the bag closed and toss it in the trash. Quick, neat, simple.

- Rubber gloves (optional).

- Gauze or disposable paper face masks (optional).

- A small portable vacuum cleaner to pick up scattered litter.

LITTER BOXES ON WHEELS?

If your litter box areas have hard floors, such as tile or linoleum, consider putting your litter boxes on wheels. You can buy large clear plastic storage boxes with built-in wheels or casters at many hardware stores and home centers. These make ideal litter boxes. Or make your own. Build simple wheeled carts using three-quarter-inch plywood (cut to the dimensions of the bottom of the litter box) and either in-line or swivel casters (available at hardware and home improvement stores). Mount the casters at the corners of the plywood and place the litter box on top. Now, when you're sweeping or vacuuming the area, you can just push each litter box out of the way with your foot! A real time-saver!

DO'S AND DON'TS FOR LITTER BOX CARE

Do scoop each box at least once a day. If you have multiple cats, more often is better.

Do use the paint scraper to scrape wet, stuck litter off the box's corners, sides, and bottom before scooping. You'll get the job done better and quicker.

Do wash your hands thoroughly immediately after doing litter box chores or handling litter care tools. Use antibacterial soap. If kids or other family members help with litter box cleaning, remind them to wash their hands, too.

Do delegate litter box cleaning to other family members if you're pregnant or immunocompromised (HIV-positive, transplant recipient, or the like). Your chances of contracting toxoplasmosis or other diseases from cat waste are very small, but not zero. If you must do the job yourself, wear rubber gloves and a face mask.

Do wear a face mask if you're allergic or sensitive to dust.

Do keep dust to a minimum by pouring fresh litter slowly, from just above the litter box.

Do clean boxes and scoops when they really need it. Scrub thoroughly with hot water. Never use ammonia-based cleaners or any other strong-smelling chemicals. Rinse, rinse, rinse, then dry and refill with four inches of litter.

Don't make the litter boxes *too* clean! The perfect litter box—from your cat's point of view—smells like *him*. It carries his comforting, familiar scent. If you throw out the unused litter too often, or scrub the box too frequently, you're destroying this very important scent environment. If you scrape and scoop the waste neatly and regularly, you'll find that you very seldom have to do a complete changeover of the litter.

To Flush or Not to Flush?

Isn't flushing used cat litter cleaner? No. Bacteria and other toxins that get into wastewater by flushing are much more likely to cause environmental problems than those that stay on dry land. Some pathogens can survive all standard sewage processing treatments, and flushed pollutants can fan out into groundwater, lakes, rivers, and the ocean.

Some recent scientific studies have shown an upswing in deaths of marine mammals due to infections with pathogens such as *Toxoplasma gondii* that are usually associated with land-based animals, particularly felines (both wild and domestic). Some researchers have speculated that the popularity in the past several years of "flushable" cat litters has caused more cat feces, and more pathogens, to enter the sewage processing stream and eventually end up in the ocean.

You'll be doing the environment and wildlife a big favor by tossing used cat litter in the trash, where it'll probably end up in a safe, modern, lined landfill. It's likely that bioremediation technologies will someday be able to effectively neutralize toxins and pathogens locked up in lined landfills.

What about composting? It sounds environmentally friendly, and many "organic," "natural," and plant-based litters claim to be "fully compostable." But even the most enthusiastic promoters of organic gardening and composting warn cat owners not to add used cat litter to their home compost piles or use it as garden fertilizer. Ideally, the heat built up in an extremely well-managed and scientifically monitored compost pile destroys the *Toxoplasma gondii* parasite that causes toxoplasmosis. But other pathogens cats can shed in their feces are more robust. And the average home gardener's compost pile, much more casually monitored, doesn't get hot enough anyway.

While the temperature of a home compost heap may be high enough in the deepest layers to kill some parasites and pathogens, it's generally too inconsistent and unreliable in killing toxo, as well as other potentially dangerous pathogens that can be found in carnivore waste.

Do make sure your cat has a box to use while you're cleaning.

Don't use plastic litter box liners. They make scooping harder, get claw-shredded, and trap wet, stinky litter underneath. And many cats dislike them.

Don't flush used litter down the toilet, especially if you have a septic system.

Don't flush used litter, even if your home is served by a city or municipal sewer system. Even if the package says "flushable," it's not a good idea.

Don't add used cat litter to your compost pile.

MOVING? BE SCENT-SITIVE

> "Watch a cat when it enters a room for the first time. It searches and smells about, it is not quiet for a moment, it trusts nothing until it has examined and made acquaintance with everything."
>
> —*Jean-Jacques Rousseau*

When you move with your cat to a new home, it's critical to establish or reinforce his good litter box habits right from the start. Though moving is a hectic time, it pays to be aware of your cat's reaction to scents he might encounter.

Moving can be very upsetting to your cat, even under the best circumstances. He'll have to get used to the sights, sounds, and smells of his new territory. To make the transition easier, provide him with "scent continuity" between his new and old, familiar environments.

- Keep your cat in a safe hideaway in your new home until the movers have finished and left. Check all doors and windows (including latches and screens) before letting your cat begin his explorations.

- Rub a slightly damp towel along your cat's back, then rub it along the floors and lower walls of your new home. Do this one room at a time as he explores each room.

- Use the same trick with a small amount of vanilla extract, or a signature perfume or scent your cat associates with you.

- Place several unlaundered T-shirts or sweatshirts carrying your scent on the floors of your new home.

- Bring along your cat's old litter box without changing the litter first, if possible.

- Think hard about where you'll be happy with litter box stations in your new home. Rather than just plunking down litter boxes anywhere, try to start out right, with litter boxes in their permanent locations.

- Same with his "old familiar" scratching post. No matter how tattered, bring it along and set it up immediately in a central location in your new home.

Be Alert for Scent Ghosts

Your cat's new territory may have been previously occupied and marked by another cat. If your cat detects "scent ghosts," old urine scent marks, he may feel compelled to overmark them, proclaiming that he's now king of this domain. Even if he's never marked or sprayed before, moving to a new home where a previous cat did could start him off on this undesirable habit.

When seeking a new home, pay attention to whether the previous owners had cats—especially if the home has new carpets. They may have been installed to replace urine-marked or stained carpets. If that's the case, and urine had soaked through to the underlayment and floorboards (and if they weren't replaced along with the carpeting and padding, as they often aren't), your cat will detect the scents, especially during humid weather, and overmark in response.

After you and your cat are moved in, let him explore his new home one room at a time. As he circumnavigates each room, watch where he stops to sniff, especially if he displays the flehmen reaction (that weird, openmouthed grimace cats make when they smell something particularly juicy). Be ready with a cleaning product designed to eliminate cat urine odors (see Resources) to clean up any areas your cat shows interest in—even if you can't smell a thing. Be sure to do a thorough cleanup. Some cat owners report that spritzing the area with a "synthetic feline facial pheromone" product (such as "Feliway") *may* help discourage further spraying or marking.

The first few days and weeks in your new home are critical. You'll be very busy and preoccupied getting settled in, but don't let a previous cat's misbehavior start up a bad habit in your cat!

Old urine scents can be much less obvious (even to your cat) during cold or dry weather, or if you're running air-conditioning. The first several months in your new abode, pay attention to your cat's sniffings around the house and be aware of anywhere he pays particular attention, especially when the weather changes or becomes humid for the first time. It's always better to clean a few spots unnecessarily than to try to correct a bad habit once it becomes established.

When Prevention Fails

Misbehavior involving urine and/or feces is fairly common. Most cats go through bouts of it from time to time. Think of it as a challenge to your detection, observation, and diagnostic skills. You and your cat will get through this together, if you remember the Three Commandments of Cat Coaching:

1. *Punishment doesn't work.* Punishment doesn't work. Punishment doesn't work. It's counterproductive. It saps your energy. It offends your cat. And did I say that it doesn't work?

2. *Don't get mad, get to work.* Getting mad at your cat, yelling and raging at him, will only confuse, frighten, and annoy him. It'll threaten your friendship and erode your bond—right when you need all the goodwill and friendship you can muster. Take immediate, positive action, but stay cheerful and upbeat. It's not your cat you're unhappy with, it's a specific behavior.

3. *To outsmart a cat, think like a cat.* No matter what *you* think about what your cat's doing, he's doing because it *works*—because it solves a problem for him. Outsmarting him means giving him one or more alternative, acceptable ways to solve that problem. If you don't offer him a better deal—an alternative that works—he'll continue to use his current solution.

Now let's get to work! "House soiling" actually takes several distinct forms:

1. Inappropriate urination (urinating outside the litter box).

2. Defecating outside the litter box.

3. Spraying (urine-marking on vertical surfaces).

4. Marking (urine-marking on horizontal surfaces).

Each has different causes and different solutions. The one thing they all have in common is that it's a *big* mistake to ignore them. And although they all involve urine or feces, only inappropriate urination is primarily elimination-related. The other three employ the cat's waste products as communication media.

Remember Mewphy's Law? Litter box problems, left to themselves, go from bad to worse. At the first sign that your cat is spraying, urine-marking, or urinating or defecating outside the litter box, you need to take *immediate* action. We'll talk about each of these steps in detail:

1. Prevent the spread and escalation of the problem.

2. Diagnose the cause. Figure out *why* your cat is doing it (which can be a combination of factors, including illness and stress).

3. Treat contributing medical causes.

Why Bathmats?

Many cats who start thinking outside the box choose a bathmat as an alternative litter box. Why?

Most bathmats (and other small area rugs) are coated on one side with a nonslip or nonskid coating, an important safety feature. The nonskid backing, though, includes chemical components that smell just like ammonia—which makes the mat smell very much like cat urine, especially when it's wet (as bathmats often are). This also applies to nonskid rug pads and mats that are designed to underlay area rugs.

If your cat has taken to using a bathmat or small area rug as a substitute litter box, keep it hung up on a towel bar or the edge of the bathtub when not in use. Or use a bathmat that does not have a nonskid coating—and be extra careful when getting out of the tub.

4. Deny access to soiled areas, and clean and deodorize thoroughly.

5. Eliminate the causes of the misbehavior.

6. Restore peace and calm.

7. Eliminate or prevent bad habits.

8. Be alert for recurrences.

PREVENT THE SPREAD AND ESCALATION OF THE PROBLEM

Delay helps to strengthen undesirable habit patterns. If your cat is ill, delay prolongs his pain and discomfort.

If you catch your cat in the act of urinating in an inappropriate location, immediately confine him in an easily monitored location (a bathroom is ideal) with a litter box. Provide as many amenities as you can on short notice: a small scratching post, comfy cat bed, a few favorite toys, and so on. This is *not* punishment. You're just trying to keep things from getting worse while you formulate a plan of action.

Identifying the culprit in urine-related misbehavior in a multicat household can be quite challenging. Use your "sixth sense," intuition, and knowledge (based on past observation) of each cat's temperament and habits. The better and more closely you've observed each cat, the better prepared you'll be to finger the offender. It's wise to discreetly monitor all cats while you search for clues. If a cat is trying to hide inappropriate elimination, you might see him dashing or slinking away from you. That's an important clue—keep an eye on that cat.

One clever technique for identifying the urine-marking culprit in a multi-cat family is a motion-sensitive video camera. Say you have a particular spot that keeps getting urine-marked, but you haven't managed to catch anyone in the act. Set up your camera in that room. You can either review the pictures later or, with some types of cameras, have the video sent directly to your web browser, smart phone, or other device. (See Resources.)

Another technique for identifying the culprit is *fluorescein dye*. Give your chief suspect a small amount of this harmless dye solution orally. (Your veterinarian can give you the dye solution and tell you how to use it. Follow instructions carefully.) In normal light, the treated cat's urine will show up as light yellow-green. But it's much easier to spot with an ultraviolet ("black light") flashlight, available at pet supply outlets.

In a dark room, under this purplish light, the treated cat's urine will start glowing a vivid, fluorescent apple green within half an hour and will continue to glow for about twenty-four hours. (Normal cat urine also shows up under black light, but it doesn't glow brightly.) If the only green glow you see is in the litter box, wait a day or so and try the dye test on your next suspect. Remember that there could be multiple culprits.

Emergency warning:
If your cat (especially if he's a male cat) cries out as he attempts to urinate; dribbles out tiny amounts of urine, in or out of the litter box, but still seems uncomfortable; or strains repeatedly but produces no urine at all, call your veterinarian immediately. Male cats who experience bladder problems can develop urinary blockages that can kill them within hours.

When you identify (or suspect) which cat has the problem, schedule an appointment with the veterinarian for a checkup as soon as possible. Inappropriate urination in particular is often caused by illness. If you suspect two or three cats but can't figure out which is the culprit, take them all to the veterinarian for a checkup.

Diagnose the Cause

Why would your perfectly behaved cat suddenly start spraying walls or urinating in the corners of the living room? Let's look at each type of elimination-related behavior problem separately.

OUTSMARTING INAPPROPRIATE URINATION

Though it's sometimes called "random wetting," to your cat, it's anything but random. If he suddenly picks a spot other than his litter box to urinate, he has a good reason—at least in his own mind. Inappropriate urination is one of the most frustrating of all cat problems. As the real reason many cats are given up to shelters (despite whatever the owner claims), it's probably one of the major causes of death in domestic cats. Adult cats with inappropriate urination problems too often find that the shelter is their last stop before euthanasia.

What might cause your cat to start urinating outside his litter box?

1. *Urinary tract infection (UTI).* For a cat with a UTI, urinating can be painful. He may associate that pain with the litter box, and avoids the box to avoid the pain. Unfortunately, his pain and distress strengthen his natural instinct to seek hidden, out-of-the-way spots to urinate. Carpet is a frequent choice.

   A cat with a urinary tract infection may or may not have blood in his urine. If you see a cat dribbling reddish or pinkish drops of urine, or find a pink stain in the litter box, suspect a UTI, and schedule a veterinarian visit as soon as possible. You don't want your cat to decide *litter box = pain*. A course of antibiotics is usually all that's needed to clear up a urinary tract infection. Follow your veterinarian's instructions carefully. Give your cat the entire prescribed amount of medication for the full length of time prescribed, even after he's feeling better.

2. *Cystitis (inflamed bladder).* A cat with cystitis has a nearly constant, uncontrollable urge to urinate. He simply can't get to the box every time he feels the need to go. He may also urinate spontaneously when he's startled, frightened, overstimulated, or excited, without being aware of it. Cats are naturally clean, fastidious animals, so he probably feels as bad about this as you do. He may hide, cry, or act out-of-sorts. Cystitis is a very complicated malady, and no one is sure exactly what causes it. But it's definitely known to be associated with stress. Cats with cystitis seem to be more sensitive to all kinds of environmental stressors, and the distress caused by the illness itself can greatly increase their stress levels. These cats are often subject to sporadic, painful flare-ups of the disease (see box on page 118). A cat with cystitis needs and deserves immediate veterinary attention. Schedule that appointment. Meanwhile, take steps to eliminate possible causes of stress in your cat's life.

3. *Other illness.* Diabetes, hyperthyroidism, cancer, neurological disease, and intestinal parasites can also cause pain and distress during urination. Litter box avoidance may be the first sign of these conditions, or of constipation, colitis, or other gastrointestinal illness. Schedule a veterinarian visit.

4. *Age-related disabilities.* Muscle pain, stiffness, arthritis, or impaired cognitive functions can make it difficult for an elderly cat to get to the box in time. He may get confused and forget where it is until it's too late.

5. *Intercat conflict.* A bully cat might stand guard at the litter box location, chasing off other cats (or one particular cat) who approaches, or intimidating or frightening a shy or low-status cat. The victim feels forced to go elsewhere, usually a well-hidden or easy-to-disguise location, like the corner of a carpet, behind furniture, or deep in a closet. Unless you witness the bully in action, it may be a long time before you realize there's a problem.

6. *Dirty litter box.*

   • You haven't kept the box clean enough.

   • Diabetes or kidney disease can cause a cat to excrete large amounts of urine, flooding the box and making it unpleasant for other cats to use.

   • Diarrhea can have similar results.

   • If too many cats are sharing too few boxes, the users may find the rapid buildup of waste with multiple scent signatures distressing or even threatening.

7. *Litter box is unattractive or unpleasant.*

   • Box is too small.

   • Covered box is trapping odors, or making users feel trapped and vulnerable.

   • Odors from harsh or smelly detergent or disinfectant are repelling cats.

8. *Litter (filler) is unpleasant or repulsive.*

   - Litter has a powerful, perfumey, or other unpleasant odor.

   - Litter has an uncomfortable or unfamiliar texture or paw-feel.

   - Sudden change in type of litter (odor, texture, and so forth).

9. *An "anti-tracking" mat is making walking around the box and jumping out of the box uncomfortable for the cat.*

10. *There isn't enough litter in the box for digging and burying.*

11. *Litter box is in an undesirable location.*

    - Noise, excessive human or feline traffic, or other distractions

    - Near a noisy, scary piece of machinery (such as a furnace that roars to life suddenly)

    - No clear visibility of approaching threats, including other cats

    - No clear escape routes in at least two directions

    - Too close to food and water bowls (cats dislike eliminating near where they eat)

12. *Litter box has been moved to a new location.*

    - Change of location was too abrupt.

    - New location is too far from old location; cat can't find box (risk for young and elderly cats).

    - New location is undesirable (see #11).

13. *Cat had unpleasant experience while in or near box.* A cat who's been punished, threatened, grabbed, jumped, surprised, or startled (by machinery, humans, or other cats) while in or near his litter box may associate the unpleasant experience with the box and take his business elsewhere.

14. *Cat avoids humans by avoiding box.* A cat who's been punished, hit, or threatened by humans *anywhere* in the house—and understandably feels he's in danger of more of the same—may decide to avoid all locations, including the litter box, where he might encounter humans. Instead, he'll head for the dark corner of a closet, where he'll feel safe.

Cystitis in Cats: A Painful Puzzle

Cystitis, a painful inflammatory condition of the feline bladder, is usually cat-
egorized as a FLUTD, or feline lower urinary tract disorder. Cystitis can be
caused by any number of bladder irritants, including crystals, infections, and
bladder stones. Often the underlying cause can't be pinned down at all, leading to a diagnosis of FIC, or feline idiopathic cystitis. "Idiopathic" simply means
"of unknown origin."

FIC (like a lot of inflammatory conditions) tends to come and go whether it's
treated or not. A cat who's experienced one bout of cystitis is 50 percent to 70
percent more likely to get it again. FIC is thought to have at least some genetic
basis. And it's almost always linked, in some way, to stress.

Cats with cystitis do have abnormalities in their bladders—but no one
knows whether the abnormalities cause the cystitis or result from it. And no
one knows whether there are similar changes in other organ systems, because
no one has really looked. The scientific studies just haven't been done.

One researcher, Dr. Tony Buffington of the Ohio State University College of Veterinary Medicine, set out to change that. He studied cystitis cats'
entire stress response systems, rather than focusing just on their bladders.
He postulates that some cats—possibly because of early life or even prenatal
events—have hypersensitive stress response systems. These cats seem to
have much less ability than normal cats to adapt to normal stressors in their
environments, such as sudden movements; unfamiliar objects, people, or cats;
loud, unexpected noises; constant noise; and a feeling of lack of control over
their environment. While most cats can usually go with the flow, a situation that
lacks predictability sends these FIC cats around the bend, greatly interfering
with their ability to cope with their environments. Dr. Buffington has coined the
term "Pandora Syndrome" to describe these cats, and he believes that FIC is
just one part of a body-wide, systemic disorder.

The FIC cat's stress system is constantly overreacting, causing signs
including misbehavior; mood changes; and a variety of "sickness behaviors,"
such as lethargy, decreased grooming, vomiting, diarrhea, pain behavior, and
decreased social interaction. All these together signal a change in the cat's
basic motivation: Rather than enjoy life and engage socially, he withdraws,

15. *Cat avoids other cats by avoiding box.* A cat who's having a territorial or
    other conflict, or who's been in a major fight, may decide to avoid loca-
    tions where he may encounter any other cats, especially bullies, rivals,
    or enemies.

conserves his energy, and takes great pains to avoid danger and upset. Feeling vulnerable all the time, he feels a strong need to retreat and recover. Essentially he goes into basic survival mode.

That's no life for a cat! And making it worse is the typical human reaction to one of the chief signs of chronic cystitis: puddles of urine all over the house. Because I have two FIC cats here on Bobcat Mountain (the lovely Angel and her equally lovely sister, Chrysanthemum), I know just how frustrating and heartbreaking it can be to see these cats suffering and to deal every day with the constant cleanup and stress.

So, is there any hope? Yes, says Dr. Buffington. First, what *doesn't* work? Scolding or punishing the cat is worse than useless. Commonly suggested remedies such as glucosamine chondroitin and cranberry supplements don't work either, for a variety of reasons. Some cat owners have reported success with drugs such as amitriptyline (Elavil) and fluoxetine (Prozac), but their efficacy in FIC cats remains scientifically unproven.

So what *does* work?

1. Feeding wet food to increase water intake.

2. Scrupulously clean water dishes. Try a large, flat water bowl, filled to the top, suggests Dr. Buffington.

3. Cat fountains.

4. Excellent litter box hygiene.

5. Extra litter boxes.

6. Environmental enrichment (toys, regular interactive playtime, climbing structures, access to window perches for bird-watching, kitty videos, etc.).

7. Reducing stress in the household. This is key.

8. Keeping the cat's environment predictable and cat-controllable.

9. L-theanine (two studies have confirmed efficacy of this supplement for FIC cats).

OUTSMARTING OUTSIDE-THE-BOX DEFECATION

This can happen for most of the same reasons as inappropriate urination. If the feces are deposited in an obscure, hidden location, such as a corner or closet, or underneath furniture, the cat may be either ill or unhappy with his litter box.

Because cat feces have little odor once they've dried, a cat can get by with this behavior for quite a while before anyone notices. Daily vigilance is key:

- If you detect the odor of cat feces in a location where there shouldn't be any, investigate. Never ignore the evidence of your nose. Keep your eyes open, especially in corners, closets, and seldom-used rooms.

- If you notice that there are no feces, or a much smaller amount than usual in the litter box, start looking around.

Your cat could be experiencing painful or difficult defecation. Keep an eye on the normal shape, size, color, appearance, and amount of his feces. Note any changes, especially if feces appear outside the litter box. Thin, pencil-width feces can indicate a partial blockage, which can be serious as well as painful. An unusually large amount of hair in feces could mean that your cat is over-grooming because of stress, pain, or itching.

The itching and aggravation of internal parasites can cause your cat to avoid his box. Look for worm segments in feces. (They look like long, skinny strands of spaghetti, or tiny grains of rice.) The rice-like objects are tapeworm segments, and they might still be wiggling.

Don't give your cat an over-the-counter dewormer. Instead, gather a fresh feces sample, pop it into a clean plastic bag, and take it to your veterinarian for testing. To prescribe an effective remedy, your veterinarian needs to identify the specific type of parasite and consider your cat's age and medical history.

Outside-the-box defecation can also be caused by territorial or status conflicts. Though cats usually instinctively bury their waste, wild felines sometimes mark prominent landmarks in their territories with scat piles—feces deposited on elevated rocks so potential competitors can both smell and see that this is already somebody's turf. This behavior is called "middening."

If little or no attempt has been made to hide the feces, the offending cat might be:

- Advertising or bragging about his status as top cat

- A "pretender to the throne," issuing a visual and olfactory challenge to the reigning top cat, or to the perceived owner of the territory

- Trying to secure more territory for himself

- Feeling insecure or threatened by a new cat in the household, and reasserting his territorial boundaries

- Reacting to the presence of a stray or neighbor cat patrolling outside

- Reacting to changes in the household, especially scent-related changes such as new carpeting or furniture, or a new person

- Reacting to seasonal changes, such as open windows in spring or summer that let in unfamiliar scents, possibly of other cats

To figure out which cat in a group has been marking with feces, some veterinarians suggest adding tiny bits of bright-colored nontoxic crayons to the diet of each suspect cat in turn. Ask your veterinarian for advice and tips before trying this.

Some cats prefer to urinate and defecate in separate boxes—another reason even single cats should have two litter boxes available. Make it easy for them. If your cat is defecating very close to or just outside the litter box, try adding another box. Pick up the misplaced feces and place them in the new box.

OUTSMARTING URINE-MARKING AND SPRAYING

Feline urine is a marvelously versatile substance. Just a few drops can deliver a whole newspaper's worth of messages to other cats. Though urine spraying and marking employ urine as their medium, neither has anything to do with elimination. Marking and spraying are *communication*, occasionally aimed at humans but usually directed at other cats. Altered and unaltered cats of both genders can (and do) spray and urine-mark.

Spraying

Spraying is the deposition of a small amount of urine, at about cat height (eight to twelve inches off the floor), onto a vertical surface, such as a wall, window, chair or table leg, door, or corner. Targets are selected for their prominent location, their importance within the cat's territory, and how readily they will be smelled and seen by other cats.

Your best bet for stopping spraying before it ever starts is to spay or neuter your cats before they're six months old. Even after spraying starts, though, altering halts the activity in 90 percent of male cats and 95 percent of female cats. But

Danger! The usual height a cat sprays is a common height for standard electrical outlets. *Urine sprayed into an outlet can cause a short circuit or electrical fire.* If you're seeing spraying in your home, protect all electrical outlets:

- Place furniture in front of outlets.

- Use child-proof outlet covers, available in the "child-proofing" section of your local hardware store or home center.

- If the outlet isn't currently needed, cover it with a plain, solid cover with no openings.

Power strips (used to plug in multiple cords for computers) can also short circuit or catch fire if sprayed. Keep them above floor level, and fitted with covers or guards. The simplest way to protect computer equipment, keyboards, and cords from a spraying cat is to deny him access to rooms where the equipment is located.

Covering up unused telephone jacks is also a wise idea.

stress, the presence of strange cats, frustration, anxiety, and abrupt changes in daily routines can trigger spraying in even the most well-behaved cat.

A spraying cat backs up to the target and raises his tail. The tail vibrates rapidly and the cat squirts urine backward onto the target. Usually very little urine is deposited. But a few drops (especially from an unaltered male cat) are pungent enough to do the job.

Why do cats spray?

1. *Social conflict or pressures.*

   - Too many cats in the household

   - Cats in the yard or just passing by

   - Loss of a cat, causing reshuffle of social group

2. *Sexual stimuli (even in altered cats).*

   - Hormonal changes in cats undergoing puberty

   - Female cat in heat

   - Sexually aggressive male

3. *Addition to the household.*

   - New cat or other pet
   - New family member (baby, spouse)

4. *Conflict/anxiety concerning a person.*

   - Reaction to physical punishment
   - Unhappiness with prolonged absence or change of schedule

5. *Conflict with another pet (usually another cat).*

   - Redirected or displaced aggression
   - Territorial conflicts

6. *Changes in routines or household scents.*

   - Moving to a new house, especially if marked by previous owner's cat
   - Remodeling: new carpeting, furniture
   - Visitors or houseguests
   - Scent of a strange cat on a visitor's clothing or shoes

7. *Boredom/frustration/general stress and anxiety.*

   - Not enough opportunity for real or mock hunting
   - No place to get away from it all

You've Got U-mail

If the intended recipient of this powerful visual and olfactory communication is an outdoor cat patrolling the yard (which even strictly indoor cats often see as part of their territories), the spray target will likely be a doorway, door, or window.

Wandering strays, NCs (Nobody's Cats), and OPCs (Other People's Cats) often spray the outsides of homes—doorways, porches, railings, foundation plantings, windowsills—with their own territorial urine marks. Indoor cats, unable to challenge or physically drive off these intruders (which human residents may be unaware of), respond with a flurry of spraying and marking of doors, windows, the corners of outside walls—anywhere outside air comes in, carrying the interloper's scent.

The presence of a sexually receptive cat outdoors can also set off a flurry of spraying and marking. Even spayed and neutered cats can respond powerfully to the scent calls broadcast by intact cats. Unaltered cats are irresistibly drawn to respond to the ancient mating call. They'll show remarkable persistence and ingenuity in escaping and accessing the object of their desires.

If your previously placid cat suddenly starts spraying, it *might* not have anything to do with your home, or his relationships with you or with your other cats or pets. He may be carrying on a heated olfactory debate with Terrible Territorial Tom from down the street or responding to the seductive scent of Flossie Fluffy in heat.

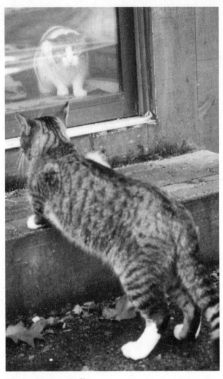

You've got u-mail!

Urine-marking

Urine-marking, like spraying, involves depositing a small amount of concentrated urine, but on a horizontal surface. The target might be a table, kitchen counter, or stovetop. Like spraying, it's usually territorial: a declaration of ownership, reassertion of boundaries, a challenge, or an attempt to reassert status and ownership in the face of a perceived threat or challenge.

Anxiety-related Spraying and Marking

Sometimes spraying or marking is focused on objects closely associated with a particular person: a favorite chair, computer keyboard, bed, pillow, purse, or clothing. Sometimes the target is the person herself. The cat may consider the target person and all closely associated objects (objects that carry that person's scent) important landmarks within his territory, and feel a strong need to remind other cats of his ownership.

Or the cat may be expressing anxiety about a perceived threat to, or change in, his relationship with the target. Maybe that person has been away from home for a while or has been working long hours. The cat might be feeling nervous about the target's relationship with another person, cat, or pet. Such emotion-based marking or spraying can show up along with other anxiety-related behaviors, such as withdrawing from company, unsociable or prickly behavior, hiding, unusual fighting or aggression, and other behaviors not typical of the cat.

If the target person (perhaps a new boyfriend or spouse) is the source of the cat's anxiety, that person should be encouraged to dispense treats, provide special attention, conduct playtimes, and feed the cat (and keep his belongings out of range until the crisis is over).

Frustration-related Spraying and Marking

Cats like things done their way. Spraying or marking can be a reaction to a cat's frustration when things aren't going his way:

- He's frustrated that his favorite person has been away too long or doesn't seem to have enough time for him.

- He's unhappy with the type or amount of food he's getting—especially a risk after a too-abrupt change in diet.

- He's restless and edgy because he's not getting enough predatory action— mock hunting or real hunting—every day.

Still mystified? Keep a behavior diary to help identify whether a particular situation or combination of circumstances might be triggering the unwanted behavior. This is especially helpful when the misbehavior shows up only sporadically or occasionally, or if you haven't been able to identify an obvious reason for it.

Keep your veterinarian in the loop. She and her staff have lots of experience with a wide variety of feline health, behavior, and misbehavior problems, and are experienced in outsmarting even the most difficult cats. Never hesitate to ask them for advice or assistance in figuring out what your cat is up to—and for tips on outsmarting him!

DENY ACCESS TO SOILED AREAS AND CLEAN UP

Once a spot has been marked with cat urine or feces, it'll continue to entice cats, both the offender and others, until it's thoroughly cleaned and deodorized. To cats, it now issues an irresistible scent call: *This is an okay place to go!*—even if you've cleaned it repeatedly and can't smell anything yourself. Even after you've completely cleaned and thoroughly deodorized a target spot, one or more cats may still be tempted to use that area simply because they remember having used it before. While your cat is relearning good habits, block access to the area, or at least discourage too-close investigation.

PREVENT FURTHER SOILING: EXCLUSION AND DETERRENCE

- Close off the soiled room for several weeks, if possible.

- Spraying a synthetic feline facial pheromone product (such as Feliway) on or near the spot *may* help deter re-marking

- Place layers of aluminum foil over the spot—cats dislike walking on foil.

- Place several solid air fresheners on the spot. Use lemon, other citrus, or a strong, perfumey scent. Cats dislike strong, sharp odors.

- Sprinkle fresh citrus rinds over the area—most cats dislike the odor of citrus. Add fresh rinds frequently, because they dry out and lose their sharp smell.

- Place a plastic carpet runner or carpet protector, spike-side up, over the area.

- Place a temporary feeding station (a bowl of dry kibble will do) near the soiled area. Cats will generally not urinate or defecate in areas where they habitually eat or snack.

- Obtain a motion detector (available at electronics and pet supply outlets) that sounds an alarm or delivers a mild electrical shock to a trespassing cat. (Before doing this, reread "'Nice' Matters" in chapter 4.)

Whatever strategy you choose, keep deterrents in place for several weeks after all cats are back to their usual well-behaved selves. Be alert for backsliding. Check the area often.

Cleaning Up

Cat urine stinks. Unless thoroughly and rigorously eradicated, it tends to return from the dead at the first hint of humid weather. You can wash, steam, or professionally shampoo the carpet so that it looks and even smells lovely to you. But unless you remove the source of the odor that calls cats back to that spot, they *will* return. Your relatively weak nose is not the one to please here. You need to convince your cat that "nothing ever happened here."

FINDING THE PROBLEM AREAS

Portable, battery-powered ultraviolet (black light) flashlights are a big help in finding cat urine hidden in carpeting, upholstery, mattresses, closets, corners, and other out-of-the way locations. Follow the directions on the flashlight package. In a very dark room, hold the flashlight close to the suspected spot. Urine, feces, vomit, and other organic material will glow faintly greenish in the light. You'll be able to clearly see the edges of a urine spot. But remember, if you're looking at carpet or upholstery, what you're seeing is just the tip of the iceberg. Much of the urine has soaked through and is hiding in the carpet backing, the padding, or even the underlying floorboards.

UNDERSTANDING CAT URINE

Because of their evolutionary origins as desert dwellers, cats don't normally consume as much water as other mammals, so their urine is much more concentrated. Cat urine is composed of a number of substances, including urea, uric acid, pheromones, sodium, creatinine, bacteria (usually five different strains), salts, and other chemicals. Because of the higher concentration of uric acid, cat urine smells worse than the urine of other mammals. As cat urine dries, the bacteria in it break down the urea, resulting in that familiar, obnoxious ammonia odor. As the biological breakdown process continues, it releases *thiols,* chemical organosulfur compounds that have unusually strong and repellent odors, and that can be detected by human noses in extremely small concentrations. So it starts out bad and gets worse.

Many of the elements of cat urine can be cleaned up with ordinary household cleaners, fooling you into thinking you've solved the problem. But the substance that causes the odor, uric acid, cannot. Uric acid can't be removed simply by washing with soap and water, or even with strong chemical cleaners. If you try this, you'll end up with an area that smells like soap-and-chemical soup—and still broadcasts *Pee here!* to your cat. Worse, some household cleaning products can actually "set" urine stains into fabrics and other surfaces. Uric acid is not water-soluble, and can't be broken down by water, soap, vinegar, baking soda, hydrogen peroxide, bleach, or other ordinary cleaners. To get rid of the uric acid and the odor, you need to use a product specifically designed to break it down or other-wise chemically or biologically eradicate it.

Hot and acidic, fresh cat urine provides a perfect breeding ground for bac-teria. It has a low surface tension, meaning it spreads and seeps rapidly into the smallest pores of every surface it touches. It seeps into cracks, soaks into fibers, and wicks under flooring. In materials like carpeting and upholstery, urine spreads out rapidly in a pyramid pattern, spreading sideways as well as down through the layers of materials.

As the urine dries, it forms uric acid salts, or crystals. These crystals have very hard shells, so dried urine is much harder to break down than fresh urine. What's more, the other chemicals in the urine form a sort of glue that helps bond the uric acid crystals tightly to whatever material they touch. Prolonged exposure to uric acid can actually "burn" surfaces and fibers, leaving dark spots and stains that cannot be removed. When dried uric acid crystals are exposed to moisture or humidity, they revive, and the smell comes back—just as the great scientist Einstein noted. If a spot has been urinated on multiple times, there may be many layers of dried uric acid crystals present, necessitating multiple applications of an effective cleaning product.

GENERAL CLEANING TIPS

There are several types of cleaning products that can be effective in breaking down uric acid and eliminating urine odor permanently, *if used correctly*. (See Resources.) Some products use specialized proteins called enzymes that break down the uric acid into ammonia and hydrogen gases, which then evaporate. Others contain ben-eficial live bacteria that actually eat and digest the odor-causing proteins. Still other products encapsulate and permanently lock in the odor-causing molecules so that they can't stink. There are also products that utilize a combination of these strate-gies. In general, no matter which type of product you use, these rules apply:

1. In order to work, the product must come in physical contact with *all* the odor-causing substance.

2. You must follow product directions *exactly*.

3. You must give the products enough "wet time" to work.

4. You must let the product air-dry.

Select an effective product. Test it in an inconspicuous spot to make sure it won't stain or discolor your carpet, upholstery, or other surface. Pay attention to any shelf-life or expiration dates noted on the container. If the product contains beneficial bacteria or other live organisms, be sure to start with a fresh supply, not the five-year-old bottle from your basement. Bacteria start to die off as soon as the product is bottled, and extreme heat or cold can kill them all, rendering the product useless.

The number-one reason cat owners have poor success with otherwise effective cat urine cleaners is that *they don't use enough.* Be generous. Make sure the product comes in contact with *all* the urine. The second problem is *not following label directions.* Each product requires a different procedure. Some require dilution; some prohibit it. Some products have to sit for a while after dilution and before use. Some enzymatic products work better when slightly warmed.

Before use, carefully read the label and any other materials that come with the product. Follow directions exactly. If you still have questions, or want additional tips and information, contact the manufacturer. You'll usually find a toll-free phone number and website on the package. Most manufacturers have loads of useful instructions and tips on their product websites, and they encourage customer calls and queries. Take advantage of their expertise. Many products come in quart or gallon jugs that can be hard to handle. Get a few empty plastic spray bottles to dilute and/or dispense the product. Look in the garden department of your local discount store or any garden center.

Some effective products need at least twenty-four hours to digest the protein in the waste and eliminate odors. If the label says to keep the area saturated and you let it dry out, you've wasted your time and money. To keep the treated area wet, cover it with a sheet of plastic tarp material or a large, sturdy garbage bag. (Don't use printed plastic; the print might transfer.) Weigh down the plastic with several old phone books or other heavy objects.

After the recommended amount of time (check the label), uncover the area and let it air-dry completely. With many products, it's a bad idea to shampoo or clean the carpet soon afterward, because the product needs to remain in the carpet to digest residual odor-causing substances.

Use your black light flashlight to check treated areas. If they still glow, you're not done yet. If the treated area has little or no green glow, it's probably okay.

What About Steam Cleaning?

Steam-cleaning appliances are a great boon for cat owners, as they allow easy, safe cleaning and sterilizing of a variety of household surfaces. *But never use steam appliances for cleaning up cat urine and cat feces, especially from carpeting.* The heat of the steam will permanently set odors and stains by bonding the odor-producing proteins with the fibers in the carpeting.

Some cat owners have reported that use of a synthetic feline facial phero-mone product (such as Feliway) on urine-marked and sprayed areas can help dis-courage re-marking. Use *after* cleaning and deodorizing the area.

CLEANING UP CAT FECES

Cat feces are much easier to clean up than cat urine, especially on carpet. Many products that remove odors also claim to remove stains. Study product labels carefully and select a product appropriate to the mess you have.

On hard, nonporous surfaces (tile, vinyl, linoleum, sealed wood), use paper towels to pick up cat feces. Dispose in the same way you do when scooping your litter boxes. Flood the area with cool, fresh water and blot dry several times. Apply an enzymatic or other cat-odor-specific cleaning product according to label directions. If the mess has left a stain, use a stain-removing product appro-priate and safe for that floor surface.

On carpeting, confine the mess to as small an area as possible. Pick up with paper towels, taking particular care to avoid rubbing, smearing, or spreading. Alternately blot with towels or rags soaked in cool, fresh water and blot dry. *Don't soak or saturate the area, or let water penetrate to the backing or padding of the carpet.* Keep wetting and blotting, wetting and blotting. A wet–dry vacuum cleaner (sometimes called a Shop-Vac) that extracts water will help immensely.

If the area is stained, apply a product designed to remove cat urine stains according to label directions. You may need to reapply it several times until the stain is completely gone.

CLEANING UP CAT URINE

Cat Urine on Hard Surfaces

If your cat's had the good grace to make his deposit on a sealed, nonporous floor—or if you've had the good sense to replace your carpets with such practical surfaces—you're in luck.

- Blot up the liquid with paper towels.

- Flood the area with cool, clear water and blot it up.

- Repeat several times.

- Clean the area with an enzymatic cleaner or other product specifically designed to eliminate cat urine odors. (See Resources.) *Follow label directions carefully.*

The Tip of the Iceberg: Cleaning Fresh Cat Urine on Carpet

Never use vinegar on fresh cat urine. It will make the situation worse.

Fresh cat urine is acidic, and relatively clean. As it dries, the urine turns into alkaline salts in which odor-causing bacteria thrive. Making the urine even more acidic by adding vinegar leads to even more alkaline salts, and more stinky bacteria. The alkaline salts then continue to absorb moisture from the air in an endless cycle, encouraging more bacterial growth. The revived odor will lure cats back to the spot, again and again. If you smell ancient cat urine on humid days, it may be because someone tried to clean up fresh cat urine with vinegar.

If you catch cat urine while it's fresh, it's easier to clean up thoroughly. As soon as you notice fresh urine on the carpet, blot up as much of the liquid as you can:

- Lay a thick layer of paper towels on the wet spot. Cover it with a thick layer of newspapers, or a couple of large old terry-cloth towels. Stand on top and step back and forth. Your weight will help the absorbent materials take up as much of the liquid as possible. Add dry paper towels if the first batch gets soaked.

- Keep doing this until the area is just barely damp.

- Rinse the area with plenty of cool, fresh water.

- Soak up the water. If you don't have a wet–dry vacuum cleaner, repeat the paper-towel steps until you've extracted several batches of clear water.

- Immediately discard the wet paper towels in an outdoor garbage can or other receptacle inaccessible to your cat.

- Besides removing as much of the cat urine as possible, you also need to remove any chemicals or cleaning substances present in the carpet (such as carpet cleaners or shampoos used previously). These chemicals and cleaners can seriously reduce the effectiveness of enzymatic cleaners. So be generous with the water.

- Clean the area with an enzymatic cleaner or other product specifically designed to eliminate cat urine odors. (See Resources.) *Follow label directions.*

Beyond the Tip: Cleaning Cat Urine Beneath Carpet

When a cat urinates on carpet, what you see and can feel on the carpet's surface is often just the tip of the iceberg. The urine has usually spread downward into the carpet backing and padding, and even onto the floorboards below, in a pyramid pattern. Those areas have to be treated, too. For best results:

- Exclude your cat from the area until you've completed the entire cleanup, and the area is clean, dry, and odor-free. This may take several days, because you may have to re-treat heavily saturated areas several times.

- Move or remove furniture on or near the wet area.

- Starting from the nearest corner or edge (most cats who urinate on carpet choose corners or edges), lift the carpet completely off the floor and fold it back. Hold down the corner with a heavy weight.

- You'll now see the wet carpet backing and padding. Remove all cat urine from these in the same way you removed it from the carpet surface. You'll probably have to lift and fold back the padding, too.

- If you see a wet spot on the floor underneath, treat that, too.

- Treat all wet areas with an enzymatic or other cat-urine-specific product, according to label directions.

- Wait until everything is completely clean and dry before putting the carpet back in place.

Cleaning Old Cat Urine on Carpet

If your cat's been urinating on your carpets for a while and you're a bit late in figuring out what's going on, your task is a bit different. You need to neutralize the alkaline salts (the bacteria-friendly dried urine) to discourage further odor-causing bacteria from growing.

- Exclude your cat from the area (from the room, if possible) for the duration of the cleanup project.

- On old urine, you need an acid solution to counteract the bacteria-friendly alkaline salts. Flood the area with a 50:50 solution of plain white vinegar and water to help bring the alkalinized urine back down to a neutral pH (acid–alkaline balance). Note: Don't use this method on fresh stains (see page 131 for why)!

- Alternately flood and blot the area with the water-vinegar solution and plenty of paper towels or old terry-cloth towels. Stand on the towels and step back and forth so they'll absorb the maximum amount of liquid. Use a wet–dry vacuum cleaner if you have one.

- You'll probably have to pull up the carpet and padding and treat the underside, as well as the floor underneath.

- Follow your vinegar-water applications with several applications of plain, cool water.

- Treat all wet areas with an enzymatic or other cat-urine-specific cleaning product.

- You may need to re-treat several times. Be suspicious, and be patient!

- Wait until everything is completely clean and dry before replacing the carpet.

If the problem is long-standing and severe, or if much of the carpet has been saturated repeatedly, your wisest course may be to admit defeat. Remove and discard the carpet, padding, and (if necessary) underlying floorboards. When selecting replacement flooring, consider an easily cleaned, hard, nonporous sealed flooring surface, such as tile or sheet vinyl.

The Carpet Conundrum

After many years of living in the company of indoor cats, I've come to view wall-to-wall carpeting as an unfortunate dead end on the road to healthy, stress-free, cat-friendly decorating nirvana. Too bad—it seemed like such a good idea.

Ever watch a cat urinating on carpet? He injects a stream of urine directly into the carpet, through the backing, and into the padding, where it spreads out several inches or more in all directions within seconds. Once that spot has been marked, it becomes a PEE HERE! sign. Before long, as the cat (and his buddies) make return visits, the padding becomes saturated, and the noxious liquid soaks into the underlayment and even the floor joists. Given the feline predilection for concealing his chosen locations for out-of-the-box urination, this can go on for weeks before you have any idea it's even going on at all.

Even without cats in your life, wall-to-wall carpeting may not be a great idea. According to a 2001 report in *The Guardian,* a British newspaper, carpets harbor toxins and pollutants at concentrations up to ten times higher than polluted city streets. One researcher discovered that the average ten-year-old carpet contains two pounds of dust laced with heavy metals (lead, cadmium, and mercury), pesticides, carcinogenic polycyclic aromatic hydrocarbons (PAHs), and polychlorinated biphenols (PCBs). Carpets and padding are also reservoirs of allergens, particulate matter, cigarette smoke, cooking residues, residues of solvents and other cleaning products, dust mites, fleas, fungi, and molds. Not exactly what you want yourself, your family, or your cats to be padding around on, bare-pawed.

The Carpet Conundrum, Solved

You've just learned how complicated and labor-intensive cleaning cat urine from carpeting really is. You've also just learned that carpeting may not be the healthiest complement to indoor living. If you plan to share your life and home with cats (or other companion animals), and if you value your own and your family's health, consider this radical suggestion: Next time you're remodeling, redecorating, replacing carpeting, or designing or building a new home, forget wall-to-wall carpet. Instead opt for sealed tile, slate, sheet vinyl, linoleum, or similar materials. Avoid flooring materials that are porous or have seams or cracks through which moisture could wick. Ask the installer to seal the edges with clear caulk. Healthier, lower stress, easier to clean—and no place to hide gallons of stealth cat urine.

Cleaning Cat Urine on Upholstery and Mattresses

This is similar to cleaning carpet, but trickier. The urine has likely soaked deep into the cushions or mattress. If you find the wet spot right away:

- Remove all washable pillows, throws, bedding, etc., and prepare for laundering with a cleaning product specifically designed for cat urine and for use in your washing machine. Follow directions carefully.

- For dry-clean-only items (bedspreads, comforters, and the like), seek out a professional cleaning establishment with experience in deodorizing cat urine on those fabrics. Ordinary dry cleaning won't remove the odor, and may permanently set it.

- Using several thick terry-cloth towels, blot up as much of the liquid as possible from the cushions or mattress. If you can, stand on the towels to help them absorb as much liquid as possible.

- As with carpet, follow with repeated applications of cool, fresh water, alternately blotting and wetting. Use a wet–dry vacuum cleaner if you have one.

- Clean the area with an enzymatic cleaner or other product specifically designed to eliminate cat urine odors. (See Resources.)

- Mattresses and cushions are generally much deeper than carpeting, so you'll need to *inject* the enzyme cleaner or other product deep into them. (Remember, enzyme and similar products must actually be in contact with the urine in order to digest the proteins and neutralize odors.) Using an injector (a tool somewhat like a turkey baster, available at home goods and hardware stores, or from professional janitorial suppliers), inject plenty of the cleaning product deep into the cushions or mattress. Follow label directions carefully, and allow plenty of time for the product to work and the area to dry thoroughly.

Cleaning Cat Urine on Walls

When a cat sprays, you'll see a vertical "dribble" of urine down the wall to the floor. On painted or sealed walls, blot up as much urine as possible, flush with plenty of cool, clear water, and apply a cat-urine-specific deodorizing product according to label directions. The urine may have seeped behind a baseboard or molding strip along the floor and onto the flooring itself. You may have to remove a section of the baseboard to remove all the urine and deodorize this area.

On wallpapered walls, you may want to cut out the affected strip of wallpaper, clean the wall underneath (if it's wet or damp), let it dry thoroughly, and replace with a fresh strip of matching wallpaper.

Outsmart Misbehavior by Eliminating the Cause

Much spraying, marking, and outside-the-box defecation and urination are caused by stress. Cats are high-strung, sensitive, territorial creatures, extraordinarily susceptible to stress. They can react powerfully and immediately to changes in their environments and to alterations in their daily routines—changes that seem positive to us or that we don't even notice. A stray cat crossing the yard can throw an indoor cat into a world of worry.

If you suspect that spraying or marking is being aggravated by the sight or smell of strange cats patrolling your neighborhood, it's wise to restrict, as much as possible, your cat's chances of seeing or smelling the interlopers:

- Close drapes.

- Make windowsills inaccessible or uncomfortable.

- Move "perch-friendly" furniture away from windows and into the center of rooms.

- Offer attractive alternatives, such as climbing trees, placed well away from windows.

- If you see your cat up on the sill, spoiling for a fight, distract him with a vigorous interactive play session or treat, well away from the window.

- Consider having a neighborly chat with the wandering cat's owner (if you know who it is).

Cats love routine and fear change. A cat's normal routine or familiar environment can be disrupted by his favorite human's burst of overtime at work, hospitalization, or vacation; a new cat in the house or neighborhood; a new baby or spouse; a strong new scent (such as new carpeting); an illness; or even a difficult veterinarian visit. The resulting stress and uncertainty can lead quickly to misbehavior, as the anxious cat tries desperately to communicate and relieve his uneasiness and restore his sense of ownership and comfort in his surroundings.

RESTORE PEACE AND CALM

The kindest and most effective strategy for dealing with a cat who's marking, spraying, urinating, or defecating outside the litter box is to remove or reduce the stress level in his environment, and restore his sense of confidence, peace, calm, and comfort. Unfortunately, the opposite usually occurs: The owner finds a pile or wet spot, flies into a rage, and swats the cat or, worse, rubs his nose in it. The rage is understandable, but this reaction does much more harm than good.

Instead, take a deep breath. Review this chapter and formulate a sensible plan of action.

Most important, don't take it out on your cat! He's doing what he's doing because it *works*—because it solves a problem for him. To outsmart him, you have to offer him at least one alternative solution that effectively solves his problem in a more acceptable way. You have to persuade him that the behavior that *you* want is actually the solution to his problem that *he* wants.

When you feel the frustration and rage bubbling up, instead of yelling at your cat, say this to him with patience and love: "It's not you, my friend, I'm unhappy with. It's just this particular behavior. We'll solve this together!"

WHAT ABOUT "SYNTHETIC FACIAL PHEROMONES?"

"Synthetic feline facial pheromones," relatively new to the market, are said to mimic the scent of the facial pheromones (found in the glands near the cheeks, lips, and foreheads) that cats habitually deposit on friendly, safe, familiar objects in their environments when they rub their faces on them. Available in both spray and plug-in diffuser form, these products are widely available through pet stores and catalogs (see Resources; one brand is "Feliway"). Although there's currently little hard scientific evidence for their effectiveness, many cat owners report success in using pheromone products in a variety of situations.

For spraying and urine-marking, thoroughly clean and deodorize the spot; then spritz the area with the pheromone spray. If the soiled area is on carpet or upholstery, clean it thoroughly; then spritz the spray on a separate piece of cloth, and lay that cloth over the deodorized area.

According to the manufacturers' claims, when the cat returns to "freshen up" his urine mark, he'll detect the scent of the pheromones instead and get the *I'm okay, you're okay* message. He'll discover that the spot doesn't really "need" urine-marking after all—it's already tagged with the universal *Everything's okay here!* scent.

Some owners also report that these products can help make it easier to persuade skittish cats to go peacefully into their carriers for trips to the veterinarian or elsewhere. Spritz the inside of the carrier about thirty minutes before the trip. Some veterinarians and clinics have also reported successful use of synthetic pheromones in calming down ill, injured, hospitalized, and stressed cats.

When your household experiences periods of unavoidable stress, such as moving, a remodeling, holidays, visits from houseguests, or other unusual events, you might want to try the plug-in form of a synthetic pheromone product. According to the manufacturers, the plug-in device diffuses a comforting, familiar

The Tragedy of Cat Hoarding

Every home has an inherent "carrying capacity" for cats. Each cat owner also has an individual and finite carrying capacity: the amount of money, space, time, emotional energy, physical abilities, and attention she has to devote to cats and their care. Some cat owners can keep large numbers of cats without ever falling into the typical downward spiral of true hoarding. A few, though, stretch their cat count past their environment's carrying capacity and discover they can't stop.

According to a June 2002 report published by the Hoarding of Animals Research Coalition, the definition of a hoarder is "someone who accumulates a large number of animals; fails to provide minimal standards of nutrition, sanitation and veterinary care; fails to act on the deteriorating condition of the animals (including disease, starvation, and even death), or the environment (severely overcrowded and unsanitary conditions); and fails to act on or recognize the negative impact of the collection on their own health and well-being."

A study by researchers at Tufts University (G. Patronek, "Hoarding of Animals: An Under-Recognized Public Health Problem in a Difficult-to-Study Population," *Public Health Reports*, January/February 1999) found that three-fourths of (identified) animal hoarders are single, divorced, or widowed females. Over half live alone. Almost half are over sixty years old, while an additional 37 percent are between forty and fifty-nine.

Researchers who study hoarding situations believe that compulsive cat hoarding may be linked to one or more of several psychological disorders, including delusional disorder, obsessive-compulsive disorder, attachment disorder, dementia, and others. Scientific evidence is hard to come by, though. What is known is that the recidivism rate for hoarders is nearly 100 percent: "Once a hoarder, always a hoarder." Hard-core hoarders are not just "normal cat lovers" who have somehow become overwhelmed; they are ill.

Most cat hoarders, though, start out as sincere, cat-loving individuals who do their best to care for the cats they take in. Often they gravitate toward rescue groups, where their growing need to acquire cats is seen as heroic rather than pathological.

Sooner or later, though, the hoarder, no matter how sincere, gets overwhelmed. Even as the hoarder's resources (time, money, space) become strained, she convinces herself that she is the only one who can properly care for her cats. Hoarders become their own worst enemies. They form extremely strong emotional attachments to their animals, often losing track of the boundaries between them and their animals. Paranoia often sets in. Family and friends who try to intervene or help out, no matter how sincerely or sensitively, quickly become ex-friends. The hoarder comes to believe that the world is out to deprive her of her precious cats. So hoarders tend to isolate themselves, and the problem worsens.

How can you tell whether someone is a cat hoarder?

1. Even though her home is already crawling with cats, a hoarder can't say no to "just one more."

2. Hoarders see themselves as "saviors" or "rescuers" and don't see any realistic limits to their "rescue mission"—space and money constraints don't seem to faze them; they believe that somehow they'll be able to save *every* cat.

3. Hoarders are unusually paranoid about what happens to animals in shelters, choosing to believe old horror stories. In fact, the care their own cats get has often become significantly inferior to the care they would receive at a shelter.

4. Hoarders frequently avoid veterinarians, either because they fear being discovered, or they believe veterinary care is actually harmful.

5. Hoarders are *extremely* resistant to re-homing any of their cats, insisting that every cat is much better off, happier, healthier, etc., with them than it would be anywhere else. If anyone offers to take one or more of their cats, they always have lots of excuses why it's absolutely impossible. Often these excuses involve accusations of wishing harm to the cats—losing yet another supporter, and further isolating the hoarder.

6. Hoarders are *extremely* resistant to having anyone visit their homes. Again, they always have lots of excuses. Even if they start out with lots of friends and family support, hard-core hoarders typically become more and more socially isolated as they push people away.

7. Hoarders often hoard other things besides cats, such as tons of old newspapers and magazines, junk, broken appliances and furniture, etc.

8. Probably the most puzzling and heartbreaking characteristic of hoarders is their inability to see what everyone around them sees so plainly: layers of filth, deplorable living conditions, ill and injured cats going untreated, even dying and dead cats. But hoarders nearly always insist that everything is fine and that their cats are healthy, happy, and well cared for.

Most hoarders aren't bad or evil, and really do want to do right by their cats. But it's the cats who suffer in the end.

scent of feline facial pheromones throughout your environment. You might also try using these products to restore peace and calm when a cat returns from the veterinarian carrying that scary "vet clinic scent," and after catfights.

PHARMACEUTICAL ASSISTANCE

Anxiety-related and territorial spraying and marking can be greatly minimized, and even eliminated, by a sensitively designed program of behavior modification, retraining, and anxiety-reducing drugs. Buspirone (BuSpar), fluoxetine (Prozac), and other prescription medications have been used successfully to reduce or eliminate spraying in cats. These drugs aren't tranquilizers. They help the cat relax while enabling him to continue relearning good habits.

After ruling out medical problems as a cause of urine-marking and spraying, ask your veterinarian about the possibility of using one of these drugs. Remember that no medication is a quick fix or cure-all. And unless you eliminate or correct the original cause of the cat's anxiety or stress, and thoroughly clean and deodorize all targets, the unwanted behavior will likely resume when the drugs are withdrawn.

LIMITING OR REDUCING YOUR CAT POPULATION

Every household has a natural "cat carrying capacity" beyond which it's unwise and unhealthy, for people and cats alike, to go. No matter how much you love cats, don't make the mistake of adopting more cats than your household resources (which include space, time, and attention) can support. Doing so may place you, one day, in the heartbreaking position of having to reduce the number of cats in your household in order to preserve an acceptable quality of life for the others, and to save your own health and sanity. (See "The Tragedy of Cat Hoarding," page 138) In households with too many cats, spraying, marking, and other difficult-to-live-with misbehaviors usually worsen until the number of cats is reduced. Don't let your household come to this sorry pass.

Outsmart Persistent Bad Habits

Once you've diagnosed the problem, cleaned up, and attacked the root causes, it's essential to take positive action to prevent the misbehavior from becoming a habit. If the misbehavior has already become a habit, take immediate action to retrain your cat and restore his usual good manners.

CAGE CONFINEMENT

Inappropriate urination or litter box avoidance can start out for perfectly under-standable reasons, such as illness, stress, or a bully cat. But if you don't solve the problem quickly, the bad behavior can persist long after the original source of stress is removed or forgotten, becoming an ingrained bad habit.

One solution to chronic litter box avoidance and inappropriate urination is *cage confinement*. It's usually very effective, but it can be rougher on you than on your cat. Still, it can mean the difference between being able to live happily with a retrained cat, and becoming so badly frustrated by a chronic inappropriate uri-nation problem that you feel your only choice is to surrender the cat to a shelter, behavior problem and all.

Cage confinement works, even for deeply ingrained litter box avoidance. It can be extremely difficult to see your cat confined for weeks to a small cage. But you and your cat will be much happier when the problem habit is gone.

1. Before beginning cage confinement, rule out illness with a complete veterinary checkup. If your cat has developed a habit of inappropriate

Cage confinement worked wonders for Chrysanthemum—but was tough on her human "mom."

urination because of a urinary tract infection or other illness, it's wise to confine him to a small room (such as a bathroom) until he's feeling better and back in the habit of using his litter box reliably. If you caught the medical problem reasonably quickly, chances are this will take only a few days. But if inappropriate urination has become a habit, a week or so of cage confinement may be the simplest way to restore and reinforce your cat's good manners.

2. From a pet supply catalog or store, obtain a cat cage large enough to hold a litter box, a comfy cat bed, and small water bowl. A cat show–size cage or large dog crate works well.

3. Place the cage in a central part of the house, where your cat will feel part of the family goings-on. Don't stash it off in a hidden corner or basement. Make sure the spot you choose isn't drafty, or too cold or warm.

4. Keep a bowl of cool, fresh water in the cage at all times. A small, tip-proof "puppy bowl" works well. Or use a water container that clips onto the bars of the cage.

5. You can either feed your cat in the cage, or let him out to eat, supervised, at his regular food location. Serve him tasty meals of his favorite foods, and then return him immediately to the cage.

6. For the first several days, keep your cat in the cage all the time, except possibly at supervised mealtimes and supervised interactive play sessions. *Warning:* This will be tough on you.

7. While he's incarcerated, talk to your cat, cheerfully and frequently. Praise his good behavior lavishly and often.

8. Keep the litter box in the cage scrupulously clean. Praise your cat whenever you see him using it.

9. Keep a positive, upbeat, cheerful attitude. Don't think of the cage confinement as punishment—it's not. It's a brief, necessary reorientation of your cat's habit patterns, a midcourse correction. The idea is to make sure your cat has *no choice* but to use the litter box when he needs to go. You need to reinforce in his mind the association between elimination activities and his litter box.

How long will it take? It depends on the cat, and on how ingrained his bad habit is. The "three-week rule" (any habit, bad or good, takes three weeks to form, and three weeks to break) is a useful rule of thumb. You know your cat and the extent of the problem best. But keep him confined for at least a week. Longer is better, if you can stand it.

10. After the first week, try graduating your cat to a larger but still confined area, such as a small bathroom, with a clean litter box, cat bed, and water.

11. The first time he urinates or defecates anywhere except in the litter box, return him to the cage for at least two days. Then repeat step #10.

12. If he uses the litter box faithfully while in the bathroom, your cat can again graduate to a larger room. Any backsliding puts him back in the cage to start over.

If you're consistent, and maintain an upbeat, positive attitude, cage confinement can work almost like magic. Even the most stubborn cat usually sees the wisdom of cleaning up his act. If your cat continues to behave at every stage, always using the litter box and nothing else, gradually give him more and more of his old freedom back. Be vigilant. He may try to test your resolve. Remember to keep praising good behavior.

SAVING YOUR POSSESSIONS AND YOUR SANITY

If you're living with a sprayer, a persistent out-of-the-box thinker, or a cat in recovery, there are a number of temporary defensive strategies you can take to outsmart some of the undesirable consequences until he's completely cleaned up his act. These simple techniques will not only preserve your possessions from damage, but also calm you down and reduce the overall stress level in your home—good for you, good for your cat.

- Remove books, papers, and other objects you value from the lower shelves of bookcases and cabinets. Remove, and store temporarily in a cat-free zone, anything that usually lives about a foot or less off the floor (within spray range).

- Keep items particularly valuable to you, whatever their location, stowed away in a cat-free location for the duration of your cat's retraining period. Your stress level will drop dramatically if you're not constantly fretting over your collection of first editions or rare LPs.

- Keep stray items, especially fabric and personal items such as towels, purses, and clothing, picked up off the floor and hung up in closets or put away. For some reason, cats having a bout of out-of-the-box thinking seem to find such objects extraordinarily appealing as targets for urine and feces. Remove temptation.

- Temporarily pick up and store throw rugs and area rugs. If they've been targets, this is a perfect time to launder them or send them out for cleaning and deodorizing. (Seek out a professional cleaner with experience in deodorizing cat urine.)

- To avoid making upholstered furniture pieces attractive targets, temporarily remove throws and pillows. A cat looking for a place to hide urine will often choose a cushion, mattress, throw, blanket, or pillow because he can direct his urine into, under, or behind it, and the urine will sink in and disappear, as it does into carpeting.

- If a particular chair has become a favorite target, cover it with a waterproof bed pad. These are quickly and easily changed and laundered. Or cover upholstered furniture with plastic furniture covers (available from home goods catalogs) covered with washable slipcovers. Then, if "mistakes are made," you need just launder the slipcover and wipe down the plastic cover.

- If your cat has shown any tendency to urinate or spray on beds, deny him access to bedrooms. If that's impossible, cover all mattresses with waterproof mattress covers (inexpensive zippered vinyl covers, widely available at home and discount stores). Use waterproof mattress pads, too. It's much easier to launder a few sheets and wipe down a plastic cover than to clean and deodorize a mattress. Cover pillows with zippered vinyl waterproof pillowcases.

- Obtain a roll of inexpensive plastic tarp (available at home goods and hardware stores) for temporary covering of furniture, floors, and so on.

- Conspicuously inspect all suspected targets frequently—*while your cat is watching you.* If he sees you frequently checking chairs and corners and sniffing pillows, he'll quickly figure out that he won't be successful hiding urine there. The idea is to reduce, to zero or near zero, the number of "safe targets" on which your cat feels he can hide urine—making the litter box seem like the preferred alternative after all.

- Close off unused or seldom-used rooms. Close closets, cabinets, and cupboards. Install child-proof latches (available at hardware stores and home centers). An area your cat can't access is one he can't urinate or defecate in, or spray or mark.

You'll be amazed how much these simple steps will reduce your stress level while your cat is cleaning up his act. But the most important thing to remember is that these strategies are all *temporary.* Your misbehaving cat can and will be retrained. You can outsmart this—together!

OUTSMART RECURRENCES WITH EVERYDAY AWARENESS

Our cats sure don't make it easy on us. By instinct and training, they're motivated to be stealthy in their elimination habits. Small wildcats carefully bury their waste to avoid attracting predators. That powerful instinct for stealth carries over to our domestic cats. Whether they're ill, and can't or don't get to the litter box in time, or they're avoiding the box because of an unpleasant association, they still have that need to hide their waste.

You're likely to see, and smell, spraying and marking quickly, since cats make little or no attempt to hide it. But inappropriate urination and outside-the-box defecation can go on for a surprisingly long time before you notice.

Get into the habit of daily, discreet observation. Sniff around. Cultivate your "everyday detective's nose." Though your sniffer will never be as keen as your cat's, by becoming aware of the clues he inevitably leaves behind (whether he intends to or not), you *can* outsmart even the stealthiest cat.

1. Make everyday observation a habit. The more you know what your cat's normal daily habits are, the easier it will be to realize when something's amiss.

2. When scooping litter boxes, be aware of the usual amount and type of waste. If you suddenly find a lot less than usual, start looking around. To be on the safe side, confine your cat to a small room with a litter box for a day or so to observe him more closely and schedule a veterinarian visit to check for medical problems. (In fact, *any* sudden or major change in the usual amount of your cat's output calls for a veterinarian visit.)

3. Has your cat been hiding in unusual locations? Watch him carefully, or confine him for a day or so. A cat who suddenly starts withdrawing or hiding may be very ill.

4. Does he scamper away when you surprise him in a corner or closet? Check that corner or closet carefully.

5. If you see a cat "scritching and scratching" at the carpet or a chair—making burying motions—investigate that area at once. Even though the offending cat is trying to hide his urine or feces, burying is a deep-seated part of the elimination ritual and sometimes the cat just does it automatically. It's an important clue. But note that another cat (not the culprit) may come along, smell the urine or feces, and instinctively try to help out by "burying" it. So don't assume the cat you see "scritching and scratching" is the culprit.

6. If you smell cat urine or cat feces, investigate. If you find anything out of place, initiate action at once. The problem won't go away by itself. It'll only get worse.

7. Let your cats help you play detective. If you see a cat paying special attention to a spot that doesn't seem like it should be that interesting, check it out carefully, especially if you see intense sniffing or flehmening (that odd, openmouthed grimace cats make when they smell something extremely interesting). Put your nose to it. If you detect any urine odor, *don't* ignore it. *Don't assume that the cat who was sniffing is the culprit.* Put all your cats under observation, and keep an open mind.

8. Ask family members for help. Has anyone observed any unusual cat behavior, sniffed any hint of out-of-place cat urine, stepped on a wet patch of carpet, seen a cat slinking into or out of a closet, or hiding in an unusual spot? Especially in multicat households, it may take the combined observation and wisdom of the whole family (including other felines) to outsmart a stealthily misbehaving cat.

9. If your cat has a history of inappropriate urination, spraying, or marking, be especially alert for any sign of unusual attention being paid to targets he's used before.

10. If your cat's misbehavior had a medical cause (cystitis, urinary tract infection), be alert for relapses. Some cats are, for unknown reasons, unusually susceptible to recurrences of these maladies.

Outsmarting Litter Box Woes: You *Can* Do It, Together

Inappropriate urination and defecation, spraying, and other chronic house soiling can be enormously frustrating, but they are solvable. Whether outsmarting your misbehaving cat is simply a matter of adding or moving litter boxes, or you have to resort to pharmaceutical assistance or cage confinement, keep a positive, cheerful, can-do attitude. Your cat will pick up your positive feelings. Once you've discovered what your cat is telling you through his misbehavior, and you've embarked on finding a solution together, you'll find that you've become closer friends and better interspecies communicators.

Quick Tips

1. Never take advantage of your cat's presence in or near the litter box to grab him and give him a pill or other medication. This is not an association you want to form.

2. Never interrupt a cat using his litter box.

3. Cats seem to take a perverse delight in deciding to use the box just as you're ready to scoop or clean it. Don't interrupt, hurry, or rush them. Be patient. Take a deep breath, sit back, and wait until he's completely finished with the entire ritual, including burying the waste, and has left the box on his own.

4. No matter how angry or frustrated you are, never swat or physically punish a cat for urinating or defecating outside his box. Never rub his nose in the mess, and never grab or manhandle him, or hurl him roughly into the litter box. (You *will* be tempted.) Any of these actions will convince the cat that you're dangerous, scary, and untrustworthy. Worse, such behavior will form or reinforce in his mind an association between his litter box and unpleasant experiences.

5. Don't make changes to a cat's litter box setup or litter or cleanup schedules without a very good reason. Make all changes slowly and gradually.

6. In all things related to his litter box, accommodate your cat's preferences to the extent humanly possible. It's worth a bit of extra effort and expense to keep him satisfied and happy in this most critical area of harmonious interspecies life.

The Nose Knows

It's astonishing how powerful and persistent an odor can emanate from a tiny drop of cat urine deposited on a countertop or windowsill. A cat who leaves such a clear, commanding message has something extremely important to say. Outsmarting him, and stopping the undesirable behavior, means listening to and decoding that message, and delivering to him a cat-appropriate, satisfying answer. It means helping him solve his problem in a way both satisfying to him and acceptable to you.

Cats don't see their urine or feces as gross or disgusting. They don't bury or hide their waste because they're ashamed of it or repulsed by it, but as a survival strategy, to avoid revealing their activities to predators and anyone else to whom they'd rather not advertise their presence.

Cats use scent as their primary means of recognizing individuals, objects, activities, opportunities, and threats in their environment. With about two hundred million odor-sensitive cells (as opposed to our paltry five million), cats live in an enveloping web of information-rich scents that we can't detect, understand, or appreciate. Subtle odors offer cats meaning, reassurance, and information. Long before computers, cats exploited their own version of the Information Superhighway—their sensitive olfactory systems—to stay tuned in to news, warnings, and local traffic, and to pursue research, entertainment, and romance.

Besides urine marks and scat piles, cats leave more subtle scent records of their passage, mood, intent, and sexual availability. With scent glands in their chins, foreheads, paw pads, and tails, they odor-mark landmarks and significant objects: their owners' legs, a doorjamb, a tree. Imperceptible to humans, these scent messages are clear as banner headlines to cats. Passing feline readers adjust their routes or activities accordingly, or superimpose their own scent messages, leaving a rich, multileveled texture of data.

To a cat, urine and feces are handy, powerful, versatile communication media. The messages in a urine mark can reveal the marker's gender, sexual availability, status, ownership of territory, and current preoccupation and agenda. It can issue a threat, sound a warning, express a complaint, mark a boundary or landmark, or just broadcast a reminder of status or ownership.

To cats, spraying and urine-marking aren't misbehaviors, but logical reactions to stress, illness, change, or a perceived threat. Punishment is both ineffective and counterproductive for litter box errors. In the cat's mind, no "error" has occurred.

The best way to minimize occurrences of inappropriate elimination, urine-marking, and spraying is to meet your cat halfway. Be a good provider, and offer him plenty of cat-friendly litter boxes that are scooped regularly but still carry his familiar, comforting self-scent.

7. Keep a "safe hideaway" ready for your cat at all times, if possible. (See chapter 4.) Or at least know what you'll need to do to put one together at a moment's notice.

8. Devise for your cat a path of least resistance to the behavior patterns you want to reinforce, while making unwanted behaviors unpleasant, inconvenient, difficult, or impossible for him.

Chapter 6

Claws and Effect

When frustrated owners complain that a cat is "scratching," they usually mean she's stropping her claws on, and shredding, furniture and other objects in their home.

Less often mentioned (because it's more annoying to the cat than to her humans) is excessive self-scratching. This almost always has a medical or psychological cause.

Let's get started on outsmarting the scratching cat by taking a look at her "little cat feet."

Your Cat's Marvelous Claws

Every feline carries an integrated system of versatile, splendidly designed, efficient tools: her paws and claws. They're spoons for scooping food; sponges for washing; cushioned, fog-light sneakers for stalking prey or silent pussyfooting; exquisitely sensitive receptors for texture, temperature, and pain; grappling hooks for climbing; clubs for swatting and stunning prey; barbed nets for trapping prey; and lethal weapons for offense, defense, and meaningful display.

Claws are vital to your cat, both physically and psychologically. As a digitigrade walker (an animal who walks on tiptoes, rather than on the flat part of her feet), she needs her claws to maintain balance and retain traction on a variety of surfaces during high-speed pursuits and playtime.

Knowing she has, always ready, a strong, well-maintained set of claws is crucial to your cat's sense of safety, security, and well-being. She's confident that she'll be able to defend herself, climb to safety, scare or chase off interlopers and competitors, and capture enough food to survive. Claws honed and ready, Mom-Cat knows she'll be able to provide for and defend her kittens.

Using the many capabilities of this marvelous tool kit is second nature to your cat. So is maintaining her claws at peak operating efficiency, ready for action at a microsecond's notice. A big part of her daily tool maintenance is stripping off old, outgrown claw sheaths and honing the freshly emerging layers to optimal condition and sharpness. Claw care is a central and deeply satisfying activity in a cat's daily life.

Okay, you might say, that makes sense for a wildcat out in the woods. But your pampered Flufferette lives in the lap of luxury, padding around on expensive carpets and snacking from a china bowl. Yet Flufferette's still a wildcat at heart. She may never need to capture live prey, escape a coyote's clutches, or chase off an intruder. But she still needs to scratch.

Scratching Is *Not* Optional!

Scratching is *not* optional. It's a normal, everyday activity, natural and necessary as breathing. Scratching serves numerous physical and psychological purposes. Your cat might scratch upon awakening from a nap; when she's feeling nervous, scared, uncertain, excited, or overstimulated; to save face after a dustup with a fellow feline; when she's bored; or when she's feeling playful and kittenish. She might scratch any old time, just for the sheer heck of it. Scratching's fun!

Wild or domestic cat, scratching's fun—and necessary.

Follow your cat around and see how often she stops for a quick scratch, or a leisurely stretch-and-scratch. If you have multiple cats, you'll notice that some like to scratch a lot, others less. Some prefer vertical surfaces; others horizontal. It's individual. But wherever and however often she indulges, a vigorous stretch-and-scratch session does wonders for your cat's physical, psychological, and emotional well-being. When she scratches, she's:

- Shedding itchy, old, outgrown claw sheaths. Her claws grow in layers. As new layers emerge, she needs to shed the old ones, like a snake shedding its outgrown skin. (The claws on a cat's hind legs also need regular maintenance, but she generally manicures these with her teeth, chewing off the outgrown claw sheaths.)

- Honing and conditioning those sharp, freshly exposed new claws.

- Getting an exhilarating workout. As she stretches her body up full length, and grips, pulls, and rakes her claws down the surface, she's stretching, exercising, and toning every muscle in her body.

- Claiming ownership of the scratched object, and marking it as an important part of her territory. Scent glands in her paw pads leave her personal scent signature on everything she scratches. You can't smell it, but to other cats, it's clear as today's newspaper: *All cats take notice: Petunia was here not too long ago, and by the way, this belongs to* her *and is an important landmark of* her *territory.*

- Reassuring herself that her defensive and offensive vital tools are ready for action, and that she's strong, fit, and able to handle whatever challenges come along.

What she's *not* doing:

- "Sharpening her claws," the more seriously to wound you when she ambushes.

- Being destructive out of spite, malice, or revenge.

- Commenting on your hopeless taste in furniture, boyfriends, or girlfriends.

- Protesting your recent frequent absences.

- Punishing you for some real or imagined slight.

So get over it—scratching has nothing to do with you. Other than its communication aspects (mainly directed at other cats, not you), it's strictly personal pampering and maintenance, like going to the gym, the manicure salon, and the therapist all at the same time.

Outsmarting Sofa Scratchers: What *Not* to Do

Your cat *is* going to scratch. If you don't provide her an irresistibly attractive scratching site that's acceptable to you, she's going to choose her own, based on her own needs and preferences.

If you catch your cat scratching something you'd rather she didn't (such as your new velvet sofa), act immediately to:

Don't want to see this? It's up to *you* to provide a superior alternative!

1. Distract her.

2. Divert her to an appealing, acceptable alternative.

3. Deter her from using the unacceptable surface in the future.

There are lots of ways to go about this. What should you *never* do?

1. Yell, scream, scold, or make a big fuss.

2. Grab her and bodily remove her from the unacceptable site.

3. Attempt to punish her in any way, including hitting or swatting.

4. Chase her away from the unacceptable site.

5. Squirt her with water if there's *any* chance she'll know *you* squirted her.

Any of these perfectly understandable (to a human) reactions will simply frighten, stress, or confuse your cat—or send her a very different message from the one you intended. They'll also create an association in your cat's mind between the unpleasant feelings of being yelled at or grabbed with *you,* rather than with the *behavior.* Because scratching is a normal, natural everyday activity—no big deal!—your cat will have no idea why you've suddenly become so upset. Worse, she might associate your sudden insane behavior with a completely different element of the situation, such as the presence of another person or cat in the room.

Instead, create an attractive diversion: Toss a small toy or treat across her field of vision. Get up and walk out of the room and call her name in a calm, ordinary voice. Get out a favorite interactive toy and let her work off some steam. Meanwhile, be formulating your deterrent-and-redirection plan. By letting you see her scratch your new sofa, your cat has given you all the tools you need to outsmart her. You now know:

- *What* she prefers to scratch

- *Where* she prefers to scratch

- *When* she prefers to scratch

With this information, you're set to make her a scratching offer she can't refuse. And if you play your cards right, she'll think it was all her idea.

Choosing Scratch Nirvana

A wildcat, or a cat who spends most or all of her time outdoors, scratches just as much as any indoor cat—if not more. But your sofa doesn't suffer, so you don't notice. Outdoor cats scratch prominent trees, fence posts, and similar structures along the perimeters of, and at important central locations within, their territories. They get all the physical and psychological

Winnie finds scratch nirvana.

benefits of scratching while leaving highly visible, scent-marked records of their presence, ownership, and movements.

If your cat spends any amount of time in your home, even just a few hours a day, she's likely to feel compelled to do some scratching indoors as well as outside. Just like indoor-only cats, she'll need a safe, attractive, easily accessible indoor scratching surface that satisfies her requirements:

- Extremely sturdy. If it tips over or (worse!) falls on her even once, she'll never return to it. (Would you?)

- Tall enough (or long enough, for horizontal scratch fans) so she can s-t-r-e-t-c-h way, way up or out, full length, and really dig in.

- Stays in one place, no matter how vigorously she tugs, pulls, and rakes it over. No slip-sliding along the floor.

- Grippy, rake-able surface, rough and tough enough to strip off her old claw sheaths without painfully catching sharp claw tips.

- Located in a central, important location within her territory, ideal as a handy signpost of territorial ownership.

Don't be surprised if this sounds like a tree. Or your sofa.

The Scratching Post

AN INVESTMENT IN YOUR CAT'S HEALTH AND HAPPINESS

Your cat's scratching post is a long-term investment, as important to her as your golf gear, bicycle, exercise machine, or other sports or hobby equipment is to you. Don't try to save a few bucks by picking up a flimsy post from a discount store. Posts and trees specifically for cats run the gamut from simple-but-sturdy to as elaborate as you can imagine. Well-designed, cat-appealing posts and trees are expensive, but worth it—for your cat's health and happiness, and for the sake of your furniture and sanity.

SAVE "OLD FAITHFUL"

Don't throw away a well-loved, well-used post when it gets messy-looking or ripped up. A heavily used, well-loved climbing tree or scratching post is a treasure.

Its condition means your cat is happy with it. Keep encouraging its use. Refresh it periodically with new catnip toys or dangly feather goodies. If it's getting so scraggly you can't bear to look at it, though . . . get a wonderful new scratching post or climbing tree and set it up next to the old one. Make the new post as attractive as possible with dangly toys and catnip. Leave Old Faithful in place until your cat's transferred her affections to the new one.

If the old post has dangly, stringy strings, clip them off for your cat's safety. If your cat chews off a stringy string and gets it in her mouth, the rearward-facing barbs on her tongue will force her to swallow it, possibly causing internal injuries or a blockage. You can also renew a favorite old scratching post by rewrapping it with fresh sisal rope, available at hardware stores and home centers.

THE PERFECT POST

In scratching tastes, as in so much else, each cat is different. Get to know your cat's preferences (vertical vs. horizontal; preferred textures). Some cats adore plain wood. Nail a three- or four-foot length of two-by-four to a wall or corner. Some cats will claim a particular length of wood trim, perhaps one surrounding a doorway they pass through often. Perhaps you could accommodate this preference? If it's an expensive milled piece, or if it would be difficult to repair or replace, this might not be acceptable. Offer an alternative she likes better.

If your cat uses a plain wood piece for scratching, she'll need a separate climbing structure, such as a floor-to-ceiling cat tree. For cats who prefer horizontal scratching surfaces, offer a selection of inexpensive corru-

Wizard *loves* his horizontal corrugated scratch pad (high-test catnip included).

gated cardboard scratch mats (available at pet supply centers). Usually treated with catnip to enhance their appeal, these mats stand up to a surprising amount of vigorous scratching. Vertical scratchers often enjoy using these mats for a change of pace.

Another option for horizontal scratchers is "grass turf" (such as Astroturf) indoor–outdoor carpeting. Attach (with glue, not nails or staples) a piece of this material to a heavy board and lay it on the floor. The texture seems to appeal to cats for both scratching and lounging.

Cats generally prefer a tougher, grippier surface than the cheap carpeting that covers most pet store and other inexpensive commercial scratching posts. (Think *tree bark*.) Most carpeting is too fuzzy and soft for a good scratch. In assessing commercial trees and posts, look for a sturdy post and a solid, stable base. If the posts come in different grades (like "standard" and "deluxe"), go for the deluxe. It will generally be sturdier and made of higher-quality materials.

If you're handy, build your own simple scratching post. Get a heavy wood beam, at least four inches by four inches or larger, and at least four feet tall (preferably taller). Mount this post *very* securely on a large, heavy base board (at least one inch thick and three feet on each side) and cover or wrap it with a variety of materials: carpet backing, sisal rope, burlap, nubbly fabric, cork panel, plain wood. Observe which scratching surfaces your cat prefers. Get as fancy as your carpentry skills allow.

Some expensive high-end climbing posts include natural tree trunks. But you can inexpensively provide your lucky cat a homegrown alternative. Bring in a tree trunk or good-size log from outdoors. Place it on the floor, or fasten it very securely to a wall or doorway. Check carefully for mold, fungus, and insects first. If you need to treat the log with an insecticide or fungicide, ask your veterinarian to recommend a cat-safe product. That log will likely get a *lot* of attention. You might find that a simple tree trunk is the easiest, cheapest way ever to outsmart a furniture-scratching cat. With millions of years of evolution on the side of the tree, your poor, ignored sofa won't stand a chance.

When choosing a post or climbing tree, keep your cat's needs and preferences uppermost in your mind. Go ahead and pick a color to match your decor (or your cat), or a "Southwest," "country," "patriotic," or "medieval castle" theme post (those, and many more, are available). Your cat won't care. But don't try to force your cat to use a scratching surface she doesn't like—not that you could. Your cat's scratching preferences are very important. Respect them.

Browse cat supply catalogs, local cat shows, and the Internet for ideas. If you have more than one cat or a particularly agile, active feline, consider a combination climbing tree/scratching post. Many of these have a spring-tension mechanism to securely wedge them between floor and ceiling. A tall post or climbing tree must be absolutely tip-proof, able to withstand lively pounces and mighty leaps without toppling.

Look for a sturdy structure with plenty of opportunities for feline fun: platforms, hideaway cave and dens, lofty sleeping hammocks. Located near a sunny window, or in a well-traveled, central part of your home, such a structure serves many purposes: bird-watching, people-watching, scrambling, climbing, working off excess energy, meditation, napping, observation of passing "prey," staging mighty pounces (real or imaginary), and, of course, scratching. Cats think and live in three dimensions. They adore heights where they can spy on whatever's going on without being noticed—or just catnap above it all.

A well-designed, well-located post will get a lot of very vigorous use, especially if you have several cats. Check the sturdiness of the joints and attachment points frequently. If the post moves around too far or too often, consider affixing it to the floor and/or ceiling with heavy-duty screws or bolts and L-shaped brackets. Even a very sturdy post can be weakened over time by lots of feline action, and a single flying leap can bring it crashing down. It could land on your cat, possibly injuring her. Or a less-than-graceful landing in an undignified pose could make her understandably hesitant to return.

If you have more than one cat, provide more than one scratching post. Cats are skilled at sharing multiple posts, but each cat will likely select a particular post as her favorite. The posts can be of different sizes and designs: some tall, some shorter; some with a variety of scratching surfaces (wood, sisal, carpet); some with cozy Berber fleece "sleeping cups" on top. Locate them in a variety of places where your cats like to hang out. Get at least one small, portable post you can move on a moment's notice to your cat's safe hideaway.

Pemi loves her scratching post with its many dangling toys.

One inexpensive option for scratching and climbing is a pair of wood step-ladders. Securely nail a broad board across the tops. Pad it with fleece or carpet, securely fastened. The board adds sturdiness and stability, and provides a great high-up observation-and-napping site. Wrap a few of the legs with sisal rope or grippy burlap fabric. Nail boards to some of the steps to make platforms; cover these with padded fleece or carpet scraps. Dangle a few feather toys. Fashion a "cave" from a cardboard box. For your cat's portable post, use a small stepladder, similarly enhanced.

LOCATION, LOCATION, LOCATION

Like a wildcat patrolling her territory, an indoor cat has daily routes and routines she follows in inspecting and patrolling her domain. Notice how she always takes the same path from her favorite sunning spot to her food bowl? Knowing your cat's daily patrol route is a vital piece of intelligence in outsmarting a sofa-scratching cat.

Get to know your cat's usual routes, and note where she habitually stops for a quick scratch after a nap, on her way to the food bowl, or after visiting the litter box. Those are the best locations for a scratching post. If your cat isn't accustomed to using a post, start off with a portable but very sturdy model, preferably including at least two kinds of scratching surfaces (sisal rope and carpet, or sisal and wood). Place it right along her patrol route. If you can, move her old target (your velvet sofa) a few feet to one side. Rub the post with some catnip. Hang a dangly feather toy.

Watch how she reacts. See which surface calls to her claws. When she scratches or scrambles up the post, praise her lavishly. Toss her a small, tasty treat. You want this post to be associated with everything good, fun, and pleasant. Most of all, you want her to stop shredding your Victorian velvet sofa. The post, if selected, placed, and introduced correctly, can be that better deal that every cat's always on the prowl for.

Outsmarting Furniture Shredders

Too many popular upholstery fabrics are perfect for scratching: nubbly-textured, satisfyingly grippy, and attached to a comfortably stable and solid object—like your Victorian sofa. Too, too tempting.

PROACTIVE STRATEGIC UPHOLSTERY

If you're selecting new furniture, or fabric for reupholstering old pieces, evaluate fabrics with cat claws in mind. Pretend you're a cat, with long, sharp claws in need of honing and polishing. With sensuous delight, run your paws over the surface. You crave a nice, rough, textured surface, good and grippy but not so loopy that it might painfully trap your claws or grab your paws enough to interrupt that gloriously satisfying downstroke.

Does the fabric you're considering seem like cat-claw heaven? Is it fair to bring such an irresistible object into your home and expect your cat *not* to sneak a few scratches? Keep looking. This is one of those little trade-offs involved in cat ownership. When you think about it, it's not really much of a sacrifice. There are likely dozens of fabrics you can choose from.

Instead of that nubbly tweed, consider smooth cotton, linen, denim, duck cloth, or another smooth fabric that won't be such a constant temptation to your cat. When your new furniture arrives, take the opportunity to renew your cat's favorite scratching post with some brand-new catnip toys.

SKIP THE LEATHER

If you intend to share your life with cats (or dogs), it seems prudent to avoid leather and vinyl upholstery. Though it's not particularly scratch-attractive, it's too easy for claws to inadvertently punch holes in it, and these can be tough to repair. The organic scent of leather may also attract undue attention from predatory critters. Stick to fabric.

SOS: SAVE OUR SOFA

Sometimes just moving an unacceptable scratching target helps. Removing an upholstered piece from a location near a window that makes it a perfect perch for bird-watching can also render it safer from the claws of your excitable little predator. Install a heated fleece kitty hammock at the window instead.

When a cat habitually uses any object for scratching, she marks it with scent glands in her paw pads. You can't smell it, but the scent will continue to draw her back. To break an unacceptable scratching habit, you need to:

1. Provide a more appealing alternative

2. Remove the attraction of the old target

Here are some tips to outsmart your cat—and save your furniture:

1. Make sure your cat has at least one sturdy, attractive scratching post, located in a prominent part of your home—somewhere she spends a lot of time, *not* stashed in a corner of the basement or the spare room upstairs.

2. Enhance the post's appeal by rubbing it with dried catnip. Hang a few feathery toys from it. Wave an interactive toy around the post. In leaping for it, your cat's claws are bound to encounter the lovely, grippy scratch surfaces. If it feels good, she'll be back for more. Make the post a wildly attractive "destination resort."

3. If possible, make the target furniture—the piece you're trying to save—temporarily inaccessible to the cat. Put it in a cat-free zone. Or turn it around or reposition it so that the target surface is no longer accessible.

4. De-scent the scratched portion of the old target with an enzyme-based cleaner, usually used for cleaning up cat urine, or a pheromone-extracting product that removes the scent that calls cats back. (See Resources.) Test the product on an inconspicuous spot first, to make sure it won't adversely affect the fabric.

5. Once your cat is reliably using her post to scratch, put the old target back in place. If the fabric is particularly cat-attractive, or if she used the old target for a long time, cover it with a smooth-fabric slipcover, at least temporarily.

6. If the furniture can't be made inaccessible during retraining, render the scratched area as claw-unfriendly as possible:

   • Wrap it with chicken wire (this will look weird, but it will slow down those claws).

   • Wrap the furniture in plastic garden netting (the kind with ½- to 1-inch-square openings).

   • Cover the site with large sheets of very rough sandpaper. (You may discover your cat *likes* to scratch sandpaper—a fact you can use in further outsmarting her. Remove the sandpaper immediately from the unacceptable scratch site and affix it firmly to her scratching post.)

- Use a product called Sticky Paws, which is much like wide, double-sided sticky tape.

- Plain double-sided sticky tape works, too, though it may leave some residue on your furniture.

- For broader areas, use contact paper, sticky side out.

- Tape some blown-up balloons to the area she's been scratching. The loud pops may spook her.

- Drape or wrap furniture with a thick sheet of plastic tarp.

- Drape the unacceptable site with plastic carpet runner, pointy side up.

- Place cotton balls soaked in a strong scent such as a flowery perfume or citrus oil on the banned furniture. Be aware that the unpleasant scent may waft over to the site you *want* your cat to scratch—and cats have extremely sensitive noses!

- Tape several pieces of fabric-softener sheets ("dryer sheets") together and drape them over the furniture. Many cats dislike the scent.

Even with these deterrents in place, your cat might simply attack another part of that same piece of furniture. You *might* get by just with using a smooth-fabric slipcover. Be vigilant, and offer an attractive alternative nearby, in a location she passes frequently.

7. Get down on your own four paws and scratch the post. You'll probably feel like an idiot, but you may intrigue your cat. (It's fun—try it!) Anything that's so attractive to you—well, she's going to want to know what *that's* all about (if only to encourage you to entertain her some more).

Some behaviorists encourage owners to carry a cat from an unacceptable scratch location when they catch her in the act to an acceptable one. I don't advise this because, in this as in all things, the idea is to outsmart the cat, not force her to see things your way. You want to persuade the cat that using the splendid new post she's discovered is *her* idea—not yours. But with enough praise and positive encouragement, it may work—depending on the cat, and your relationship.

Training the Cat

Newlyweds Bonnie and Larry set up housekeeping with Bonnie's four-year-old cat, Mickey, and a lovely new tweed-and-velvet living room suite. Mickey was in cat-scratch heaven. "Don't worry," Larry told his wife. "I know how to train cats."

For the next few weeks, every time Mickey scratched the sofa, Larry picked him up and gently carried him outside. Lesson learned. For the rest of Mickey's long life, every time he wanted to go outdoors, he simply scratched the sofa.

Defensive Decorating

Finished wood and painted surfaces, rugs, and other household surfaces can also be negatively affected by your cat's attention. When remodeling, acquiring new furniture, or refinishing old pieces, keep these defensive decorating tips in mind:

- Top wood furniture such as dining room tables, coffee tables, and credenzas with ¼-inch-thick sheets of tempered glass. This glass can be custom-cut, and the edges finished and rounded, for a surprisingly low cost.

- To protect your carpets from hairballs and the inevitable puddles of cat vomit, use inexpensive, washable throw rugs and runners. These can add style and color to your home, and they let you change your decor with your mood or the season.

- Just about everyone has some possessions that are so fragile, valuable, or precious that the very thought of their coming to harm or being destroyed can cause stress. Don't place temptation in the way of your cat, and don't drive yourself crazy with worry whenever you can't be there to guard your treasures. Instead, designate one or more "cat-free zones" in your home. These can be anything from a small closet in your apartment to an entire wing of your mansion. To preserve heating and air-conditioning circulation, consider installing screen doors.

- Consider sacrificing a piece of furniture. Perhaps that chair the cat had chosen to scratch was never really a favorite anyway? Might you be able to just leave it in place and let your cat have at it? If you must move it to a less conspicuous location, make the move gradually, just a few inches a day, until it's where you want it. But don't get any ideas about stashing it in the basement. Remember, location of scratch sites is critical.

OUTSMARTING SPEAKER SCRAPERS

Cat claws and stereo speakers seem to have a magnetic attraction, causing much household friction and static, not to mention painful woofing and tweeting from the speakers' owner. What can you do to outsmart your speaker-scraping kitty?

- Invest in speakers with metal, plastic, or aluminum coverings. These look high tech and have much less claw appeal than fabric.

- Mount speakers high on walls, on shelves or brackets well out of cat-leaping range. (Watch out for handy launch pads.)

- Place speakers on tall, narrow pedestals or shelves that don't have enough room for a leaping cat to land. Make sure they can't be easily tipped over in a desperate feline attempt at access.

- Invest in flat-panel, wall-mountable speakers.

- If you're stuck with old-fashioned cloth-covered speakers, wrap them with chicken wire, hardware cloth, or other claw-unfriendly material, much as you'd cover claw-attractive upholstery. You can spray-paint the wire covering to match the speaker's cloth and disguise your tactics.

And remember: A philosophical acceptance of a bit of damage to your possessions over time is part of the equation of living with cats. But they're worth it.

Is Declawing the Answer? *No!*

Cats come with claws—it's part of the package. A cat's natural balance, grace, and agility are dependent in large part on her retractile claws. Claws are so important to a cat's physical health and sense of well-being that it's astonishing, and horrifying, that anyone would think to deprive her of them just to spare the sofa.

Declawing (onychectomy) is controversial and serious surgery. Too many uninformed people, including cat owners, think of declawing as a quick, easy shortcut to a perfectly behaved cat. It's also misnamed. It should be called de-toeing.

De-toeing is a series of irreversible radical *amputations*. For each toe—all ten (usually) on the cat's front paws—the claw, the cells at the claw's base that allow it to grow, and the terminal toe bone are surgically removed. All associated tendons are severed. It's the equivalent of having all of your fingers amputated down to the first knuckle. Usually only the front paws are declawed. But that's more than enough to maim a cat for life and set off a cascading chain of chronic physical, psychological, emotional, and behavioral problems.

Although it can reduce scratching damage, this "convenience surgery" often leads to much more serious, permanent damage to the owner's household and to the human–feline relationship. Declawing deprives the cat of important tools and weapons she needs every day.

Declawing is complex, painful surgery. Many things can go wrong. If the surgeon positions his cutting instrument improperly, he risks removing too much of the toe, including the toe pad. If not enough tissue and bone are removed, misshapen, deformed toes can grow back, causing additional pain and disability. Tiny bone fragments left in the tissues can cause infections, which often require additional surgery—with more pain and recovery time. Abnormal growth in severed nerve endings can cause the cat to experience painful twinges and other uncomfortable sensations long after the operation.

After surgery, the cat's forepaws are wrapped in bandages to control the often considerable bleeding. If they're too tight, circulation to the feet can be cut off, leading to permanent damage. Even if all goes well, recovery can take several weeks. Many cats aren't given adequate pain relief, either in the immediate post-surgical period or during their lengthy recovery.

While the cat recovers, paws swathed in bandages, she can't dig normally to bury her waste. Soft shredded newspaper or a similar material must be substituted. This radical change alone (cats hate change, especially in their personal daily routines) can set off chronic litter box avoidance problems. In pain, unable to dig or bury, the cat may associate her pain with the litter box. (Review chapter 5 for what that can mean.)

Between the trauma of losing their claws and the horror, pain, and trauma of recovery, it's not surprising that many cats experience both acute and chronic physical, psychological, and emotional problems after this mutilating surgery. Many declawed cats undergo personality changes, becoming unfriendly and skittish, withdrawing from company and hiding away. Deprived of their claws, their natural first line of defense, many overcompensate by becoming fearful, nervous biters.

The constant stress of feeling helpless, clumsy, and defenseless can weaken a cat's immune system over time, leaving her more vulnerable to illness. Chronic stress and insecurity can lead to such maladies as cystitis, skin disorders, and obsessive-compulsive behavior patterns like self-mutilation.

Declawing sometimes causes secondary contracture of the tendons, making walking very uncomfortable. With the joints of her forepaws gone, she compensates by shifting more of her weight to her hind legs, throwing her seriously off balance, possibly causing the muscles of her front legs to atrophy. Being constantly off balance is very distressing to the cat, psychologically and physically.

Without claws, she can't scratch an itch or properly groom herself, leading to a downward spiral of poor coat condition, chronic mats, depression, and illness. Though cats also use their teeth for grooming, claws are essential in grooming parts of the body that can't be reached with nibbling teeth: the head, mouth, neck, face, and ears. Cats also use their claws to detangle fur and remove loose hairs and foreign material.

A declawed cat can't climb effectively or safely. But since she still has an instinctive urge to climb, she'll still attempt it, risking serious injury in slips and falls. Without the accurate control claws provide, even a short leap to a windowsill can prove dangerous.

Anecdotal evidence collected by cat experts and behaviorists indicates that declawed cats are much more likely to develop litter box avoidance problems, chronic chewing and biting fetishes, and other long-term serious behavior problems. Too many declawed cats, after not magically turning into the perfect pets their owners expected, are banished to basements, surrendered to shelters, or just abandoned. Declawed cats have lost tools they need for survival, so abandonment means being near defenseless. They can't hunt or capture prey. The "quick fix" becomes the cat's death sentence.

No declawed cat should *ever* be allowed outdoors. She can't climb or scramble to escape danger or a predator, she can't fight off an attacker, and she can't defend herself.

Please don't de-toe your cat! Some observers feel that declawing should be permitted when the only alternative for a "destructive" cat is euthanasia. Others feel that this approach constitutes emotional blackmail—and that anyone who would threaten to kill a cat if they couldn't declaw has no business owning a cat. But even those who feel that declawing is sometimes justified say that declawing shouldn't be considered except as an extreme last resort, when all behavior modification and environmental solutions have failed.

It's time cat owners in the United States followed the lead of most of the rest of the civilized world (including the UK and most of western Europe, as well as Australia and New Zealand), where declawing cats is either illegal or considered extremely inhumane and performed only in extreme circumstances.

WHAT ABOUT TENDONECTOMY?

A less radical surgical procedure, *tendonectomy*, is often proposed as a more humane alternative to declawing. But this, too, is serious, disabling surgery. The surgeon severs the tendons that attach the claw to the terminal toe bone, but the claw remains in place. The cat can no longer extend her claws. It's less painful and recovery is quicker, but it can lead to such problems as joint fusion and chronic, crippling arthritis. After tendonectomy, regular claw clipping is a must—a procedure cat owners hoping for a "quick, easy fix" may be unwilling to perform.

Don't declaw your cat—outsmart her instead!

Clipping Claws: Is It Compulsory?

You don't *have* to clip your cat's claws. If she goes outdoors, she needs her claws sharp and ready for defense and escape, and her everyday activities and

Jesse prefers to manage her own claw-grooming.

lifestyle will ensure that her claws stay honed to perfection. But even if she lives indoors, as long as she's active and has attractive and appropriate scratching surfaces, she'll keep her claws honed and conditioned herself. Since cats, unlike dogs, retract their claws under most conditions, your cat is unlikely to damage your floors, furniture, and other household items just going about her everyday business.

Consider allowing your indoor cat to not only keep her claws, but keep them the way she likes them: sharp and ready for action. It's one way to help offset some of the disadvantages of the decision you've made on her behalf—safety indoors vs. a risky freedom to come and go.

WHEN YOU SHOULD CLIP CLAWS

If you have young children, a frail elderly person, or an immunocompromised person living in your household, it's prudent to trim your cat's claws regularly. A startled cat could inadvertently cause serious injury to a vulnerable person.

Kittens in the clumsy-toddler stage tend to get hung up on whatever they dig into, risking injury. Don't leave them hanging. Clip until they're old enough to manage claw maintenance themselves. But accustom *every* kitten to having her paws handled frequently and claws clipped occasionally. Many veterinarians request that patients' claws be clipped before appointments.

Unable to perform the vigorous stropping and scratching needed to keep their claws honed, elderly cats are at risk of having their claws grow long and raggedy, or curve inward and grow into their paw pads, causing pain and infection. If you have a sedentary elderly cat, inspect her claws frequently and clip as necessary. No matter what your cat's age, keep an eye on her claws. If they've grown long and especially if they're tightly curved, clip them.

READY TO CLIP?

Ask your veterinarian or a veterinary technician to show you how to clip claws. With practice, it's a quick, simple procedure. It doesn't need to be done very often—maybe every two to six weeks, depending on how fast the claws grow and how sharp they're getting. Unless your cat is having problems with her rear claws or isn't maintaining them herself, you don't need to trim them. She hones these with her teeth and doesn't keep them as sharp as her front claws.

Ideally, you started in kittenhood, getting your cat used to the idea of having her paws and toes handled. Before actually trying to clip, gradually work up to

Press paw, *gently*

You need:
1. *Bright* light
2. *Sharp* clippers, sized for cats
3. *Positive* attitude

Don't nick the quick

Clip

Clipping your cat's claws

pressing lightly on each toe to "pop out" the claw. Make sure your cat is completely comfortable with this routine. Stroke her paw pads, express your admiration for the sharpness and beauty of her claws, and then pop them out to admire them all the better. Your cat will likely eat up this extra praise and attention. Remember, one easy way to outsmart your cat and get her to cooperate with whatever is on *your* agenda is to offer her as much lavish praise as you can dish out. Graciously accepting accolades is *always* on her agenda.

Pick a comfortable spot with excellent lighting. Get your tools ready: clippers and a styptic pencil (available at drugstores) or some cornstarch (from your kitchen cabinet). If you do the job right, you won't need the styptic pencil or the cornstarch—but have them ready in case you nick the "quick," the tiny pink vein that runs through the claw. If you nick the sensitive quick, it'll bleed and your cat will feel a twinge of pain.

Professional groomers prefer guillotine-style clippers, which clip the claw cleanly with a single blade. Slip the clipper's opening over your cat's claw, position it exactly where you want to clip, and squeeze. The blade slides across the opening, safely making a straight, neat cut. Another popular style of clipper works like a pair of scissors, with opposing blades. Whichever style you choose, get one sized for cats. Most dog-claw clippers are too big.

The only other things you need are a calm, relaxed cat, and a positive, confident attitude. If you fumble and hesitate, your cat will pick up on your nervousness and become twitchy herself. If she seems uneasy, try wrapping her in a fluffy towel. But if she's that nervous, the session probably won't be successful anyway. Try again later when you're both feeling calmer.

It's not necessary to clip all the claws in one session. If you get one or two done at a time, fine, especially at first. As your cat becomes accustomed to this odd routine, she'll probably let you do several at a time. Don't rush her.

Using a guillotine-style clipper, here's how to do a successful clip job:

1. Position your cat comfortably on your lap, facing away from you. Make sure you can see her paws clearly.

2. Hold your clippers in one hand.

3. With your other hand, gently press one of your cat's toes between your thumb and forefinger. The claw should pop right out.

4. Make sure you have a clear, well-lighted view. Observe where the slightly darker, pink vein (the "quick") is. You don't want to cut this!

5. Slip the clipper into place, well ahead of the quick. The clear part of the claw you want to trim, called the *cuticle,* is just a very tiny sliver.

6. Squeeze the clipper to make the cut.

7. Oops! Did you nick the quick? If you did, you'll know it! Your cat will jump and yell. Quickly dab the claw with the styptic pencil or a tiny bit of cornstarch to stanch the bleeding. Apologize profusely. And it's probably a good idea to terminate the clipping session.

8. Cat still calm and relaxed? Move on to the next claw.

9. Stop the session as soon as your cat gets twitchy or nervous.

10. All done with no nicks? Praise your cat for her patience and forbearance.

11. Follow up your session with a treat, and plenty of praise.

With claw clipping, practice makes purr-fect.

CLAW CAPS

Vinyl claw caps, widely available in pet stores, catalogs, and veterinary clinics, come in a range of sizes and a rainbow of colors. Their rounded edges protect skin and furniture from claw damage. Your veterinarian or technician can show you how to properly glue them over each claw with a special fast-drying adhesive. They must be replaced every four to six weeks, as the claws grow out. Some cats hardly notice the caps; others tear them right off (they're nontoxic). Respect your cat's preferences.

Outsmarting Itchy Cats

The first thing you notice is tufts of fur blanketing the carpet. Your cat is twitching and itching, rippling her back muscles rapidly, dashing about and stopping every few seconds to frantically scratch at her almost-naked tummy. What's going on?

Much attention is given to alleviating the miseries of humans who suffer allergies to their cats. (See chapter 11 for tips on outsmarting humans' cat allergies.) But cats suffer from allergies, too. Human allergies generally provoke red, watery eyes and lots of sneezing and sniffling. In cats, allergies more often cause skin problems: intense itching, crusty scabs, hair loss, and oozing sores from endless scratching. Cats with food allergies are often afflicted with chronic diarrhea and vomiting, which can be severe and debilitating.

Allergies in cats, like those in humans, are caused by a poorly understood overreaction of the immune system to particular substances. They can't be cured, though the sufferer can sometimes be desensitized through a lengthy immunotherapy program. Because immunotherapy takes so long, veterinarians often prescribe antihistamines, essential fatty acid supplements, or corticosteroids during the program to control itchiness. Using these medications doesn't seem to interfere with the success of immunotherapy.

Cats can be allergic to just about anything. Determining exactly what a suffering cat is allergic to can be a long, frustrating process. Once you find the allergen that's plaguing your cat, minimize exposure. Many lifestyle changes that will minimize your cat's suffering are also healthy choices for you, your family, and the environment.

If your cat suddenly develops the typical symptoms of allergy—itchiness, frantic scratching, scabby skin—schedule a trip to the veterinarian to rule out underlying medical conditions. Then start your detective work.

Is it flea season? A major cause of feline allergies is *flea allergy dermatitis,* an extreme sensitivity to a protein in flea saliva. Just one flea bite can set off a raging attack. Even if your cat stays strictly indoors, check her carefully for fleas. Using a flea comb (a comb with narrowly spaced teeth), comb your cat's fur, especially along her back. Did you find black flecks (like grains of pepper) on the comb? Shake the comb onto a dampened paper towel. If the black flecks "bleed," leaving pink stains, your cat's hosting fleas. The flecks are flea feces, and the pink stain is your cat's blood.

Look for signs of *miliary dermatitis*: small, itchy red bumps, often found along your cat's backbone. These, too, can be caused by flea allergy.

OUTSMARTING FLEAS

In the past, flea control was a hit-or-miss affair, depending on a motley combination of shampoos, powders, sprays, and collars. These were somewhat effective but had many drawbacks, including toxicity.

A more recent innovation, anti-flea pills, deliver medication into a cat's bloodstream—but each flea must bite the cat to be exposed to the insecticide. That can be a lot of bites. Some pills don't kill adult fleas, but instead prevent them from reproducing—not a quick solution for an itchy cat.

Some owners still resort to folk remedies such as garlic, brewer's yeast, or vitamin B. But there has never been any hard scientific evidence that these substances banish fleas. Worse, some traditional flea remedies, such as oil of pennyroyal and tea tree oil, are toxic to cats, and can sicken and even kill them. Oil of pennyroyal is a potent liver toxin. Never use these oils, or products containing them, on your cat.

Today the most popular flea treatments are "spot-on" medications. They've made ineffective, old-fashioned remedies like flea collars and dips obsolete. Just squeeze a drop or two of the medication on your cat's skin, along the back of her neck, to control fleas for a month or longer. Ask your veterinarian about the right anti-flea product for your cat. The dose must be adjusted for your cat's weight.

Spot-on medications mix with your cat's skin oils, which then migrate with gravity and ordinary movements. Some products flow into your cat's sebaceous glands (under the skin, attached to hair follicles), where they remain over time. Other stay on the skin surface. Because they're on or under the skin, these new medications are less likely to wash off. The spot treatments kill virtually all fleas on your cat within eighteen hours and prevent fleas from laying eggs. Spot-on products that kill adult fleas on cats include:

- Advantage (imidocloprid)

- Frontline TopSpot (fipronil)

- Revolution (selamectin), which also prevents flea eggs from hatching and controls heartworms, ear mites, ticks, roundworms, and hookworms

If your cat is allergic to fleas, a product that kills adult fleas will give her the quickest possible relief. There's also an oral medication, called Capstar (nitenpyram), that kills adult fleas.

Warning: Beware of over-the-counter and supermarket spot-on flea medications. Many of these contain permethrin, which is highly toxic to cats. Never use flea control products packaged or labeled for dogs. Always ask your veterinarian's advice before using *any* anti-flea treatment on your cat.

Another modern product, Program (lufenuron), is a flea birth control pill. It kills eggs and larval fleas. You give it to your cat monthly in pill or liquid form.

De-flea-ing Your Home

Studies show that products like Program, used together with flea adulticides like Advantage, Frontline, and Revolution, can eliminate flea infestations, even if the house or environment is not treated. This is welcome news, because treating your whole house and yard can be difficult, expensive, and environmentally undesirable. If you have to treat your house or yard, ask your veterinarian to suggest cat-safe insecticides. Keep in mind that cats who go outdoors may keep bringing more fleas home.

If your cat's having flea troubles, vacuum frequently. Pay special attention to cracks, crevices, and corners where fleas lurk. Dispose of the bag outside immediately. Wash cat bedding and all towels, throws, rugs, and other objects she lounges on in hot water at least weekly. Use a flea comb to remove remaining fleas from her coat.

OUTSMARTING ENVIRONMENTAL IRRITANTS

Like people, cats can be sensitive to pollen, mold, and dust. They live closer to the floor, where cleaning substances, particulate matter, and heavy fumes collect, so they absorb a heftier dose of toxins and allergens than we do. Wall-to-wall carpeting is notorious for harboring large quantities of particulate matter, cigarette smoke, cooking residues, residues of solvents and other cleaning products, dust mites, fleas, fungi, and molds—all substances that can trigger and aggravate allergies.

To the extent possible, eliminate these irritants from your environment. Clean and vacuum regularly, wash bedding in hot water, and use a HEPA (high-efficiency particulate air) filter. If your cat goes outdoors, keep her in during the worst of pollen season.

A surprising number of cats are allergic to plastic. Telltale signs include *miliary dermatitis* (small, itchy red bumps), hair loss, scratching, or crusty patches on the cat's face, neck, and throat—from leaning over plastic food and water dishes. Replace plastic food and water bowls with stainless steel, US-made ceramic (glazes on foreign ceramics often contain lead), or heavy glass. If your cat habitually lounges on plastic furniture or other plastic objects, drape her favorite spots securely with thick, soft towels or blankets to prevent skin contact.

Cat itchy, but no fleas in sight? Have you installed new carpeting (which often outgasses irritating chemical fumes for months), recently cleaned your rugs, or used a carpet freshener? Many common household cleaners and products, especially those that aerosolize or linger on surfaces, can cause allergic reactions, including itching and excessive shedding. Keep cats out of recently treated areas. If that's not possible, put down clean towels or throw rugs where she can lounge, and watch her carefully for signs of distress or allergic reaction.

Better still, avoid using chemical cleaners and products. Clean and sanitize your home with a steam-cleaning appliance. (See Resources.)

If you smoke, consider quitting. Besides eliminating another irritant from your cat's environment, you might be saving her (and yourself) from cancer. Research at Tufts University found a link between secondhand smoke and feline lymphoma. Even after adjusting for age and other factors, the researchers discovered that cats living in homes with smokers, and exposed to secondhand smoke, are more than twice as likely as other cats to get this deadly disease. Cats exposed to smoke for five or more years had more than triple the risk. And in a two-smoker household, the risk went up by a factor of four. Cats inhale the secondhand smoke and also consume carcinogens when they groom themselves and lick the particulate matter off their fur.

OUTSMARTING FOOD ALLERGIES

Food allergies, often signaled by both diarrhea and itchiness, especially around the face and ears, are toughest to track down. If you've recently changed your cat's diet, go back to the old food and see if symptoms subside. It may take two weeks or more for itching and other symptoms to settle down, even after the allergen is removed. If your cat's on a prescription diet, ask your veterinarian to prescribe an alternative formula.

Still baffled? The only really successful method for identifying the source of a food allergy is a strictly monitored elimination diet of foods previously unfamiliar to your cat; you then introduce other foods, one at a time, and see which cause allergic reactions. Consult your veterinarian for advice on selecting the initial diet and carrying out the program. This can be a lengthy and difficult process—no unauthorized treats, no table scraps, no exceptions. Offer plenty of extra cuddling and love during this difficult time. Frequent play sessions will help keep your cat toned and tuned.

IS IT YOU?

But what if your cat's allergic to *you*? Research at Britain's Edinburgh University suggests that some cats are allergic to their owners. The allergy is a reaction to human skin or to dust mites. These microscopic pests feed on the tiny flakes of skin we all shed naturally. Many humans develop severe allergic reactions to dust mites, so it's not surprising cats would, too. Symptoms in cats include itchiness on the face and paws, and chewing and licking large quantities of fur.

Dust mites live by the millions in mattresses, carpets, upholstered furniture, drapes, and pillows, so minimize the presence of these in your home, especially in bedrooms. Cover mattresses and pillows with mite-proof covers. Launder bedding and other washables such as throw rugs frequently in hot water. Use a laundry additive formulated to kill dust mites. Vacuum carpets, drapes, and upholstered furniture frequently, and dispose of the bag promptly.

If it's not dust mites, your cat could be allergic to a cosmetic, soap, or fragrance you use rather than to you. Minimize or eliminate your use of scented products, change brands, or seek out hypoallergenic products. You and your cat may both benefit.

Chapter 7

Fractious Felines: Fighting, Aggression, Ambushes, and Biting

Why Do Cats Fight?

Cats fight for a lot of the same reasons people fight: space, food, prime denning spots, access to members of the opposite sex, or just plain crankiness and boredom. Sometimes cats "fight" just for the sheer heck of it—for a rousing, vigorous workout. But they also "fight" because they're cats, obeying hardwired, instinctual programs laid down by millions of years of evolution. They "fight"—actually, react instinctually with an instant defensive-aggressive posture—in response to stimuli that they perceive as potentially threatening.

Cats don't intellectualize, plot revenge, hold elaborate long-term grudges, or battle to defend religious dogmas, philosophical positions, or political opinions. They *react*. A cat reacts to an immediate stimulus in his environment: another cat who tries to grab his sunny spot, an intruder cat sauntering shamelessly through the yard, being grabbed by a sticky-handed toddler, a sudden or loud noise, a bright light in his eyes. A dozing cat who's suddenly startled by a loud clamor (phone, doorbell, a yell, loud music suddenly starting up) is instantly wide awake, alert, and primed for flight, defense, or, if all else fails, offense. It's hardwired survival instinct. It's the essential inner wildcat, ready for action.

If another cat is the stimulus—or just happens to be within paw range—he's going to be the target. That's instinct, too. He's just in the wrong place at the wrong time.

It's what happens next—or what *doesn't* happen next—that's most useful and enlightening in trying to outsmart fractious felines. Most "catfights," especially among groups of altered, indoor cats who know one another well, are brief, if intense little tiffs that blow over in seconds—*unless somebody adds fuel to the fire.*

How can you add fuel to a catfight fire?

- By paying attention to it

- By rewarding it with your startled, appalled, distressed reaction

- By trying to "save" one combatant from the other

- By trying to break it up

- By screeching, throwing pillows, tossing cups of water, squirting a squirt gun

When most cat owners hear fighting, their first impulse is to rush to the scene and try to break things up. Once you do that, though, you might as well sell tickets—because your cats are going to put on a show.

The impressive sound effects—hissing, growling, yowling, screeching—can make little tiffs and brief tussles sound like serious business. (Some cats even stage fake fights to get attention, or an early breakfast.) But most of those scary sounds are just expressions of surprise or annoyance, or elaborate bluffs, challenges, and dares.

WHO STARTED IT?

You just sauntered into your living room to find two cats in a feline standoff, growling, posturing, and spoiling for a fight. Okay, guys, who started it?

The instigator is the one staring directly at the other cat. His pupils are narrowed to slits, and his whole body is facing directly forward, toward the object of his displeasure. The defender (or victim), on the other paw, has averted his eyes from the instigator. His pupils are round balls, indicating fear and uncertainty. His head is facing his opponent, but his body is facing sideways. He's poised to escape at his first opportunity.

Male cats fight mainly over access to receptive females (if not neutered), status, rank, and supremacy. Females are more likely to tussle over space and territory.

CONTENTMENT OR CONTENTION?

Like a bobcat or cougar, your cat has a hair-trigger response to sudden stimuli— he's primed to, in a nanosecond, flee or fight for his life. But he's also, by nature (his genetic heritage) and nurture (his upbringing and experiences), *neotenized*— a permanent kitten.

This dual nature means that your cat *initially* reacts to a stimulus just as an adult wildcat would—as if he were reacting to a dire threat. *Initially.* That's hard-wired instinct at work. But his next thought is likely to be either, *Oh boy, a fun game!* or *Oops!* Quite often a startled or surprised cat who lashes out at another cat will immediately realize his mistake and back off, perhaps elaborately washing a paw or grooming his perfect tail to save face. "No harm, no foul." Both cats forget the whole incident and go about their business.

But let's say instead of this happy ending to a simple, everyday misunderstanding, a shrieking human had sped into the room, wielding a squirt gun or tossing pillows or cups of water. Now, instead of two calm, face-saving felines, we have two wet, stressed, pillow-pummeled, seriously annoyed felines. And other cats are galloping in to join the festivities and choose up sides. Is this what you really want?

Escalating Conflicts

Long-term, low-level conflicts between cats can get out of hand before you realize it, and they can explode into major battles without warning. Be alert for determined, persistent aggression or repeated fights that continue, escalating in intensity and frequency, day after day. Pay attention to the belligerents' usual habits. If they're not eating or drinking normally, or if they're urine-marking and spraying all over the house or avoiding the litter box, it's time to step in. All that stress every day, as well as medical problems from the disruption of normal activities, can bring on illness and long-term behavior problems.

NEUTER AND SPAY, IT'S THE SENSIBLE WAY

One way to virtually guarantee that all brief tiffs will turn into serious tussles is to fail to spay and neuter your cats. Intact cats are, in many ways, more like adult wildcats than neotenized domestic cats. Sex is serious business and calls for intense competition and life-or-death fighting. The physiological pressures of sex hormones ramp up unaltered cats' awareness and prickliness about territory, status, rank, and access to cats of the opposite sex (especially females in heat). Male cats will engage in continuous, vicious, and even lethal fights to gain access to reproductively receptive females. If you keep intact cats in close quarters, you're asking for trouble—not to mention stratospheric veterinarian bills.

Bona Fide Fights

Cats are strong and well armed enough to kill or seriously injure one another. There are times when a wise owner steps in and safely separates the combatants. Be alert for signs of earnest fighting, and put a stop to it before things get out of hand. Use your common sense, your knowledge of your cats, and your intuition to outsmart fighting felines. If the dispute is serious enough, and the cats feel that enough is at stake, they won't just "work things out"—they'll continue and escalate their battles.

How do you tell the difference between a particularly exuberant wrestling-and-play session or brief spat of annoyance, and a serious, for-real fight? If the answer to any of the following questions is yes, it's probably time to intervene:

1. *Participants:*

   - Are either of the combatants intact (not spayed or neutered) cats? Fights between intact cats tend to be more serious and more likely to result in injuries.

   - Is one (or both) of the participants a feral or stray, or a cat of unknown origin, health, or vaccination status?

   - Is the fight between two cats who have fought seriously before, especially recently?

2. *Duration:* Has the tiff gone on more than a few minutes? Do the combatants back off, only to continually reengage at the same or higher level of intensity?

3. *Vocalizations:* Are the cats screeching and wailing loudly and menacingly? Listen for low, guttural growls, and watch for drooling, exaggerated swallowing, and licking of chops.

4. *Weaponry:* Are the combatants' claws fully extended? Are they biting?

5. *Damage:* Are large tufts of fur flying? Do you see any blood or wounds on either of the cats?

If the fighters are your own cats who're usually good friends, it's less likely to be a serious tussle than if one of the combatants is an unaltered, bad-tempered, territorial neighborhood tom. But if you think the fight is serious, or if any of the combatants is wounded or bleeding, take immediate steps to settle things down and seek medical attention for the victim(s).

OUTSMARTING COMBATING CATS

Fighting cats are highly aroused—seriously overstimulated. Even your own dear pet who sleeps with you and kisses you on the lips can become a slashing, raving stranger in seconds during a fight. *Don't* try to pick him up. Instead, try to distract one or both cats. Interrupt the proceedings long enough for one or both cats to flee. Cats prefer flight over fight—*if* they have the option. Outsmart them by distracting them, defusing the situation, and giving them a chance to beat feet.

1. Distract one of the cats by tossing a toy or treat past him. If your cats recognize the characteristic rattle of a treat can, try that. Try to make the distraction or diversion a pleasant one, if at all possible.

2. Grab an interactive toy and try to interest one or both of the cats in some vigorous predatory play. Transform their aggression into predatory aggression, targeted at the toy instead of each other. Use the toy strategically to keep them away from each other, or use a fishing-pole toy in each hand to keep them focused on their own "prey."

3. If transforming the fight into play doesn't seem to be working, or if it's critical to separate the combatants immediately (for example, if one is injured):

   - Make a loud, sharp sound: yell, clap your hands, blow a whistle or air horn.

   - Grab a handy object (such as a magazine) and fan one of the combatants with a breeze of cool air. This will tickle his whiskers, distract him, and interrupt his train of thought. Use this "cool breezing" technique anytime you need to distract a cat quickly.

   - Toss a blanket, jacket, or other cloth item over one (but not both) of the cats.

   - If they're wrapped too tightly to cover just one of them, that's when tossing a soft pillow or cup of water can be a lifesaver. The distraction will likely cause them to separate momentarily. Take advantage of the break to isolate one cat and give the other a chance to flee.

- If your cat is fighting with an intruding NC (Nobody's Cat—a feral or stray) or OPC (Other People's Cat) outdoors, and nothing else has worked, turn the garden hose on 'em. Try to maneuver your cat indoors and into a confined, safe hideaway as soon as possible. Here he can settle down so you can observe him for possible injuries.

For indoor battles, once the cats have separated (one or both will generally flee as soon as an opportunity arises), try to make sure they stay separated for several hours, until they're completely settled down. Discreetly close the door to the room where the fight took place and walk out, leaving one cat inside and one cat outside the room. If you plan to leave him there for more than a few minutes, provide a litter box, a bowl of dry kibble, and a bowl of fresh water. Playing a radio or CD with soothing music may help the combatants settle down faster. Avoid doing anything noisy or anything that might further overstimulate them.

Some cat owners have reported that using synthetic feline facial pheromones (such as Feliway) can help after fights. After separating the combatants, spritz a bit of it into the air over their heads.

Making Up

When the combatants seem back to normal, reintroduce them under pleasant circumstances—at dinnertime, for example, featuring a particularly tasty entrée. Watch discreetly for any signs of lingering hostility. Maintain normal schedules and routines; conduct playtimes and other rituals as usual. If you feel uncomfortable about their moods or intentions, escort one of the cats to his safe hideaway for an overnight time-out. Repeat the reintroduction in the morning. This is *not* punishment—just proactive prevention.

Your cheerful, positive, upbeat attitude will help immensely in reestablishing normalcy. Don't punish, yell, or otherwise make a big deal out of the fight. Punishment will mystify, confuse, and annoy the cats. Worse, it can make them more fearful, stressed, and overstimulated—and more likely to pick further fights.

After a while, the cats will be unlikely to recall exactly what it was they were so excited about. But if the fight was serious enough, they may remember: *I have a problem with that &^%$# cat.* Keep an eye on both ex-combatants for several days until you're sure they've mended fences and moved on.

In the most serious cases, you may have to proceed as if one of the cats is truly a stranger in the household, and reintroduce him very slowly and gradually, just as you'd introduce a new adult cat, complete with:

- Use of your safe hideaway to isolate the cats in rotation
- Scent-mingling techniques
- *Lots* of time and patience

Preventing Fights

It's easier and wiser to keep potentially fractious felines, and prickly ex-combatants, calm and contented than to deal with the stress and emotional disruption of fights.

PLEASED TO MEET YOU?

The arrival of a new cat in your home can cause extreme territorial anxiety and reactions ranging from sulking and hiding to all-out war. To your cat, an unfamiliar member of his species encroaching upon his territory is a potential competitor for his resources and, in his wildcat mind, a possible threat to his survival. Though adult cats usually accept kittens more readily than other adults, conflicts can arise. To minimize stress, introduce new cats and kittens very slowly and gradually, taking your cues from the cats.

- Make sure all cats—new and resident—are spayed or neutered.

- Prepare a safe hideaway for the new arrival with litter box, scratching post, food, water, and toys.

- Before bringing any cat into your home, make sure he's healthy and free of internal and external parasites. A complete veterinary checkup, including tests for such contagious diseases as feline leukemia, is a must.

- Keep your new cat isolated from other cats for as long as your veterinarian recommends. Don't cheat on this—it can affect the health of all your cats. Don't allow under-the-door "footsies" until you get a medical all-clear; block the space with a board or rolled towel.

- *Resist the temptation* to show the new arrival to his feline housemates immediately. Many experts recommend that a neutral third party (a friend or neighbor) transport the newcomer into your home, directly into his safe hideaway.

- Once your veterinarian gives the new cat a medical all-clear, start a daily bedding swap between new arrival and resident to accustom them to one another's scents. Swap a litter box, too.

- Try synthetic feline facial pheromones, spritzed on all the cats' bedding; it *may* help smooth introductions.

- To speed scent mingling, gently rub all cats with a towel dampened very slightly with well-diluted vanilla extract, or—even better—with a scent (perfume, lotion) your cats associate with you.

- The new arrival and resident cats will soon be making their own introductions—scents, vocals, and footsies—under the door. Let them.

- If possible, use a room with a screen door as your safe hideaway. After a day or so of bedding swaps and footsies, let the cats interact through the screen.

- Observe behavior discreetly. Does their body language show curiosity and friendliness, or

Pleased to meet you!

hostility and aggression? Be patient. Some cats take longer to come around than others. Let them set the pace.

- When you consistently see more curiosity than hostility, try a trial face-to-face meeting. There'll probably be some warning hisses and a bit of posturing at first. This is normal.

- Make these first getting-to-know-you sessions short and pleasant. Everybody gets tasty treats and lots of lavish praise.

SCENT MINGLING MENDS FENCES

Your cat returns from the veterinarian, groomer, or boarding facility after a short visit. His fellow felines—his friends—shun him or even attack him. What's going on?

They simply don't recognize him. Cats recognize other individuals primarily by *scent*. Perhaps he doesn't smell like himself because of some treatment or procedure he had. Or maybe he carries odors from the clinic that the other cats recognize, strongly dislike, and consider threatening or frightening. Help mend fences by restoring him to the family scent group as soon as possible.

- Temporarily escort him to his safe hideaway.

- Install a litter box that has been lightly used by the other cats.

- Bring in cat beds, towels, mats, or other items the cats habitually lounge on, as well as an unlaundered article of your own clothing.

- To speed up this scent mingling, gently rub all the cats with a towel dampened very slightly with well-diluted vanilla extract, or with a scent (perfume, lotion) your cats associate with you. The sooner all the cats share a common scent again, the quicker they'll be back on good terms.

Reintroduce the cats for a pleasant group experience, such as a regular mealtime at which you serve a particularly tasty special treat. Discreetly watch for signs of aggression or trouble. Be vigilant, and patient, and peace and order will soon return.

WARM SUNSHINE: THE NATURAL CAT TRANQUILIZER

Are your cats prone to sporadic fractiousness? One of the smartest things you can do—besides mostly ignoring their noisy bluffs and challenges—is to provide as many warm sunny spots as possible throughout your home. There's something about the warmth and brightness of *direct sunshine* that has an almost magical mellowing effect on even the feistiest feline. It seems to be a natural cat tranquilizer. Consider removing drapes, curtains, shades, and blinds to let more direct sunshine indoors.

PHEROMONE MAGIC?

Unavoidable changes or stressors, the introduction of new people or pets, or a recent serious fight can leave cats feeling prickly, stressed, uncomfort-

Tabigail is both warmed and calmed by sunshine.

able, and fractious. These difficult situations are usually temporary. But if the situation is unaddressed, their unhappiness can lead to long-term behavior problems.

When facing such a frustrating behavioral situation, it can be tempting to try something new, even if unproven. One example of such an unproven, but potentially promising, behavioral modification tool is "synthetic feline facial pheromones." Although there's no hard scientific evidence for the effectiveness of these products, some cat owners have reported success in reducing spraying and urine marking, as well as calming skittish, stressed cats in a variety of situations. It can't hurt, and it may be worth a try!

These products (such as "Comfort Zone with Feliway") are said to mimic the scent of feline facial pheromones (found in the glands near cats' cheeks, lips, and foreheads) that they habitually rub onto friendly, safe, familiar objects. To a cat, according to the manufacturers, these products say: *Everything here is fine, I'm okay, you're okay, don't worry, be happy!*

PHARMACEUTICAL HELP

Sometimes two cats can develop a deep animosity or rivalry, and continue covert and overt hostilities despite your best intentions. Even so, it's very rare that formerly friendly cats become so seriously and permanently estranged that they can no longer live together. If your cats don't seem willing to forgive and forget, ask your veterinarian about drug therapies. A temporary course of a behavior modification drug can help cats over the hump as they become reacquainted. Drug therapy is most successful when used in conjunction with a sensitively designed behavior modification program. Your veterinarian can refer you to a qualified animal behaviorist who can evaluate your situation and suggest a plan of action.

Behaviorists have noticed that it's often more helpful to medicate the victim rather than the attacker. The victim's fearful, apprehensive behavior can trigger an attack, so making the victim more assertive or aggressive can lessen the chances of attack. The drug buspirone (BuSpar) sometimes causes an increase in the aggression level of the treated cat. When he responds to attacks with confidence and even aggression rather than fear, the attacker backs off, and often changes his opinion of the target.

Selective serotonin reuptake inhibitors (SSRIs) such as fluoxetine (Prozac) and paroxetine (Paxil) are also sometimes prescribed to help in controlling aggression between cats.

Understanding and Outsmarting Aggression

All catfights have a lot in common, but a variety of triggers can set them off. The more you know about the stimuli that get cats battling, the better prepared you'll be to outsmart the combatants and restore peace and order.

TERRITORIAL AGGRESSION: TURF BATTLES

Cats are naturally much more territorial than other domestic animals, such as dogs. And females can be just as territorial, if not more so, than males. Territorial aggression is almost always directed against other cats, not people. You're *part* of your cat's territory, not a competitor for it. Your cat might challenge another cat over access to that very important, moving, can opener–operating landmark in his territory—you.

Indoor cats give up much of the territory they'd likely be able to claim if they had unescorted outdoor access. Your cat's wild counterparts patrol considerable tracts of real estate: Males generally patrol about 150 acres, females, about 15 acres. But the "core range" of a typical domestic cat is much smaller: about 100 yards in diameter. A house cat's core range often centers on the room where his

A little territorial tiff over turf atop a bookshelf.

preferred person spends the most time. Cats are at their prickliest when protecting and defending their core ranges.

An average-size home with three exclusively indoor cats has the equivalent of a density of about thirty thousand to forty thousand cats per square mile. No wonder turf problems crop up. Keep this in mind when determining your home's cat carrying capacity.

Outsmarting Turf Battles

- Think 3-D—height. Provide plenty of cat trees and climbing structures with high retreats, lofty secluded napping spots, and climbing opportunities.

- Offer plenty of cozy hiding and denning spots. Don't force cats to share, unless it's their idea.

- Set up at least one litter box per cat, plus one extra, split between at least two different stations in your home.

- Provide at least one food bowl per cat and, if possible, one water bowl per cat. Never force cats to share a food bowl.

- Introduce new residents, especially new cats, very gradually.

- Don't ignore Old Faithful in favor of that cute, fluffy new kitten.

- When life brings inevitable changes, strive to minimize their effects on your cat. Keep routines and schedules stable and familiar.

- When planning changes (a move, a marriage, a new baby), keep your cat's needs in mind, and plan for them.

- Be a good provider. Make sure each cat feels secure that there will always be enough of everything—food, attention, playtime, laps, warm sunny spots, toys, litter boxes—for him.

A cat displaying territorial aggression probably feels compelled to defend your entire home from whatever he thinks is the threat. So one of the kindest (and smartest) things you can do for him is make his task easier: Give him a smaller territory to protect. Escort him to his safe hideaway (see chapter 4), equipped with everything he needs: food, water, litter box, toys, a comfy bed. This is just the "chill pill" some territorially sensitive cats need. Play a radio softly to mask unfamiliar voices. Make this space your cat's territory—his core range—and respect it!

PREDATORY AGGRESSION

A cat who ambushes toes, wraps himself around ankles and digs in, leaps onto the shoulders of unsuspecting passersby, or stalks, chases, and pounces on other cats or pets needs more daily outlets for his natural predatory drive. Predatory aggression is often a sign of plain old boredom—and frustration. Your cat needs to chase, stalk, pounce, and "kill" every day. It's an important part of what makes a cat, a cat.

Fortunately, your cat doesn't have to kill and eat live prey to satisfy this deep compulsion. All that chasing, stalking, and pouncing is hardwired instinct. But both the procedure for administering the "killing bite" and knowing that prey is food are learned from MomCat during early kittenhood. Kittens with mothers who themselves never learned to hunt almost never perfect their kill-and-eat-prey skills. But they're still deeply compelled to go through the rituals of the chase and capture. They still thrive on the thrill of the hunt.

Vigorous, interactive play—mock hunting—is crucial for your cat's physical, psychological, and emotional health. It tones his muscles and

Bird- and chipmunk-watching is a great group activity.

heart, provides an aerobic workout, helps sharpen his senses, and offers him a deep satisfaction he can obtain in no other way. It's fun, it's healthy—it's *necessary*. Midnight ambushes, pounces from the top of the kitty tree, and toe attacks are pleas: *More play, please!*

Outsmarting Predatory Aggression

- Redirect predatory stalking by tossing a Ping-Pong ball or mini Wiffle ball past the cat's field of vision.

- Visually separate the predator from his "prey." Get between them, crouch down, then distract one or the other and divert him to a different activity.

- Set up a bird feeder just outside a favorite window perch.

- Play a "kitty video" featuring fluttering birds or scampering rodents.

- Keep a toy box in a convenient location (near your cat's favorite napping spot), filled with tempting solo toys. Include small furry

Silver exercises his inner wildcat in predatory play.

fake mice, crinkle-balls, feather dusters (known here as "big bad birds"), Ping-Pong balls, superballs, and twisted pipe cleaners. Rotate the toys periodically.

- Hide attractive toys in unexpected locations so cats can discover the prey for themselves.

- Try a solo-interactive toy, such as the Panic Mouse. (See Resources.) Some cats love this; it leaves others cold.

Simple Gifts?

If your cat has unsupervised access to the out-of-doors, his predatory aggression will take a more natural turn: He'll stalk, hunt, and probably kill small animals and, occasionally, birds. *You can't teach your cat not to hunt or not to kill birds any more than you can teach him not to breathe.* He may never be an efficient killer, and he may show no interest in eating his kills. But he'll still hunt. He's just being a cat.

Petunia presents a love token.

Your hunter may bring you sweet little gifts: small

critters, alive, dead, even partially consumed. Some cats leave their offerings on your doorstep; others prefer your pillow. Never punish your cat for this. Instead, accept the gift as the token of esteem it is, thank your benefactor profusely, and discreetly dispose of the gift. If you'd rather not find mouse parts on your pillow or dead robins on your doorstep, consider treating your cat to a safe, enriched, indoor-only lifestyle.

PLAYFUL AGGRESSION

One of the delightful traits of our neotenized cats is their lifelong love of play. Because they're predators, though, feline play often has an aggressive, hunting-like quality. This isn't misbehavior or naughtiness—it's normal. Playful aggression isn't usually accompanied by growls, spits, and hisses—just lots of high-pitched, kittenish mewing and squeaking.

As Tabigail and Silver pounce and wrestle, they're practicing vital hunting, offensive, and defensive tactics.

Kittens often see humans as littermates or playmates, fair game for playful aggression. Many kitten owners let their tiny fluffballs pounce on their toes, wrap themselves around their ankles, and run up their blue-jeaned legs . . . *so* cute! When Tiny Fluffball grows into fifteen-pound Big Bruiser . . . not so cute.

If playful aggression is directed at you (or another person), *don't* playfully swat the cat back, join in the "game," or let him continue. Instead, get up quietly and leave the area. This is *not* behavior you want to encourage. By leaving, you remove both the audience and the target.

A playful cat is a happy cat. A cat who plays every day stays healthier, lives longer, and is a much more agreeable companion than a sad, neglected cat who's been punished and yelled at so often for doing what comes naturally that he's given up. The cat who rolls and tumbles with his littermates, stalks your pen as it moves across the page, bats at the computer cursor, or leaps in wild abandon when you wave a feather toy is expressing the power and joy of his inner wildcat.

Glory in his nature with him. Remove dangers and roadblocks from his environment; provide plenty of appealing, cat-safe toys, interactive play opportunities, and climbing furniture; channel his playful energies into acceptable directions—and let the good times roll.

DEFENSIVE AGGRESSION

A defensively aggressive cat feels he's under a threat from which he can't, at the moment, escape. He wants to make that frightening or threatening stimulus or situation disappear. He crouches low and plasters his ears and whiskers flat back against his head. He hisses and spits. His hair stands up on end, rough and spiky. *I'd rather run,* he's saying, *but if you come any closer, you're toast!* He's not going to attack—unless he's attacked, *or interprets your movements or actions, or the movements and actions of a nearby cat, as an attack.*

Leave him alone until he calms down. If you can, remove whatever has aroused him. If it's another cat, try to distract one or both cats, or temporarily divert them to another activity. Toss a few treats in opposite directions and try to maneuver the cats into separate rooms.

Don't pick up a defensively aggressive cat. Don't approach him, even with soothing words and good intentions, or try to calm him down. He's scared, aroused, overstimulated, and potentially dangerous, even to someone he knows well. It might take several hours, even overnight, for him to settle down enough to eat, play, and interact with other cats normally. Don't rush him. Pay attention to his body language. Don't interact with him until he's back to his usual calm self.

STATUS-RELATED AGGRESSION AND BULLYING

A dominant or particularly assertive cat might start to take advantage of shyness or fearfulness in a feline housemate. If he succeeds in intimidating his first target, he'll escalate his aggression, and may go on to pick on other cats, too. Bully cats torment their victims in numerous ways: hostile ambushes, litter box guarding, food bowl guarding, chasing victims away from beds and dens, and just plain everyday harassment.

Although top cats in a group tend to be benevolent despots, a few succumb to the temptation to throw their weight around. A bully cat is not necessarily a group's dominant cat, but he might be making a bid for the job.

The victim cat may become deferential, fearful, and constantly nervous, always looking over his shoulder. This reaction often eggs the bully on, triggering even more aggression and harassment. In extreme cases, a cat can become a pariah within the group—low cat on the totem pole, fair game for ill treatment from everyone.

Outsmarting the Bully

When you have a bully cat, you have to outsmart both the bully and his victim:

- If your bully is guarding resources (food, litter boxes), provide multiples, in multiple locations. He can't be everywhere all the time.

- Feed Bully separately, perhaps in a small bathroom or safe hideaway room.

- A persistent and recalcitrant bully might benefit from a course of confinement to a cage or small room. (See chapter 5.) Restrict the bully's freedom and return it to him in increments, as he demonstrates his improved attitude.

- Resist your temptation to "save" or "protect" the victim. Unless the fighting is extremely serious and extended, meddling will do more harm than good.

- If you're truly worried about the situation, separate the principals and (after complete veterinary checkups) reintroduce them slowly and gradually, just as you would a new cat into your home.

If bullying or other aggressive behavior shows up suddenly, or is not typical of the cat, schedule a veterinarian visit immediately. If medical causes are ruled out, a short course of a behavior modification drug may get Bully back on track.

FEARFUL AGGRESSION

Too many changes, too fast. New people. New scents. New carpeting. A whole new house. Any of these can trigger extreme stress and fear in a vulnerable cat. Fear can work both ways: The fearful cat's shrinking-violet behavior might provoke a bully cat into attacking him, or the fearful cat may lash out himself. When changes loom, plan for and minimize their effects, and let your cat get used to them at his own pace. Provide a safe hideaway, cozy private dens, and other shelters to which he can escape when things get too frantic. Think 3-D: Lofty perches are much appreciated by fearful cats.

Cats who've had little or no experience with people other than their own humans are sometimes terrified of strangers. They might dash off and hide (the usual response) or become aggressive toward people they don't know well. Cats who were abused or mistreated can also react aggressively to strangers. (Can you blame them?) For similar reasons, some cats are fine with one gender, but fearful or aggressive with the other.

If you know your cat has a problem with strangers, especially if he tends to lash out at them, escort him to his safe hideaway well before visitors are expected. Play soft music to mask their voices.

Outsmarting Fearful Aggression

To help make sure your cat won't attack dinner guests, or the plumber, use this program of gradual desensitization:

- Start with very short visits from cat-savvy friends who know your cat is fearful and won't do anything to frighten him or aggravate his fears. Avoid loud talk and sudden movements. Avoid visits from children; kids' high, piercing voices and sudden movements seem to bring out the worst in fearful cats.

- Encourage your visitors to sit on the floor or crouch down, where they'll seem smaller and less threatening.

- Encourage your visitors to offer your cat a few treats, or toss a favorite toy gently past him.

- Never force your cat to interact with anyone. Let him make the moves.

- At the first hint of stress or fearful reaction, end the visit and reward your cat's patience and forbearance with a few tasty treats and a play session with a favorite toy.

- If your cat seems willing, gradually build up to longer visits.

- Be patient and persistent. Let your cat set the pace.

A cat whose fear is focused on another cat can be outsmarted, too, through a program of limited freedom (for the aggressor) and safe desensitization (for the fearful cat):

- Escort the aggressor (or bully) into a safe hideaway.

- At mealtimes, carry the aggressor to your feeding location in a cage or carrier.

- Feed the aggressor and the fearful cat in the same location—but keep the aggressor in his carrier.

- After meals, return the aggressor to his safe hideaway.

- Over time, let the fearful cat become accustomed to the presence of the feared cat in nonthreatening situations, and for short periods.

- Watch for signs of hostility or aggression in the aggressor. If he shows he can be in the fearful cat's presence consistently without attacking, growling, hissing, or posturing aggressively, gradually give him his freedom back. Any signs of backsliding? Back into solitary.

Another approach is to separate the fearful cat from the group by installing him in the hideaway. Then reintroduce him into your group just as if he were a new arrival in the family, taking care to maintain a gradual, cat-driven pace and making lavish use of scent mingling.

PAIN-BASED AGGRESSION

If any cat suddenly starts showing extremely aggressive behavior not typical of him, or if several incidents of fractious behavior involving that cat occur within a short time, he may be ill or in pain. Schedule a veterinarian visit without delay.

Because of their nature and evolutionary heritage, cats are extremely reluctant to show pain or reveal that they're injured. A cat who's hurting because of an illness, tumor, infection, abscess, mouth or tooth pain, or other malady, or who's been injured, can unexpectedly lash out.

Even the most tranquil and beloved cat can become a dangerous stranger when he's in pain. If your cat is injured or in pain, and you need to get him out of immediate danger or transport him for medical care, use extreme caution in approaching and handling him. Fear, stress, shock, and pain can make him lash out. If he injures you badly enough, you won't be able to care for him or get him the care he needs. Wear heavy gloves and a jacket, and keep the cat well away from your face and head. Use a large towel, blanket, or other thick material when picking up and holding the cat.

SEXUAL AGGRESSION

The surest cure for sexual aggression in cats is spaying and neutering. Unless your cat is part of a responsible, well-managed, and well-designed breeding program being carried out for the good of the breed, there's absolutely no excuse to *not* have him altered (*spayed* for females, or *neutered* for males).

Sexual aggression is seldom a problem in spayed and neutered cats. But because a cat's sexual identity resides as much in his brain as in his sexual organs, he can retain a measure of sexual drive, and a predilection for engaging in mating-like behavior, even after he's been neutered. Occasionally a neutered male cat will continue to mount females (even spayed females), although he obviously can't follow through. Some particularly confused males even mount stuffed animals. Depending upon your attitude, this can be either amusing or embarrassing.

The female object of the confused male's amorous advances might react with anything from bemusement to fond tolerance to anger. This kind of behavior is often a purely adolescent phenomenon that disappears as the cat matures. (It can recur if the neutered male encounters an intact female, though.) It can also result from boredom and lack of opportunities for predatory play and daily, vigorous exercise.

If the quasi-sexual behavior—pursuits, mounting, and nape grabbing—persists, the unwilling target female's patience can wear thin and a real rift can grow between your cats, leading to potentially serious behavior problems.

What Part of "Neutered" Don't You Understand?

If your neutered male persists in mounting and treading a female, take him to the veterinarian. It's possible that he was incompletely neutered. A simple blood test can check levels of testosterone (the male sex hormone) circulating in his blood. Sometimes further surgery is needed to complete the neutering process and extinguish the behavior.

In the meantime, make sure female cats have plenty of spots, preferably up high, to hide from

Patient Tabigail, perplexed Wizard ("I seem to be missing something . . .").

unwanted amorous advances. If you see neutered Lothario eyeing the girl of his dreams, distract him with a play session to work off some of that excess energy.

AGGRESSION FOR NO APPARENT REASON

Sometimes cats who've lived harmoniously together their whole lives suddenly decide they can't stand each other. (Sometimes people do this, too.) If this happens to your cats, look for possible causes:

- Change in family (human or feline) composition: death; new baby, spouse, or pet; a youngster who has left for college

- Loss of a protective or "therapist cat" who was protecting one cat from the covert aggressive intentions of another

- Major changes in the household: new carpeting, construction, or demolition

- Major cat-specific changes: new type of food or litter, litter boxes moved

- Move to new quarters, with lots of new scents and other stimuli to deal with

- Illness or pain in one of the cats

Start with a complete medical checkup for both cats, especially if the aggression appeared suddenly or has escalated rapidly. If medical causes are ruled out, put on your Sherlock Holmes detective cap and figure out what's bothering your cat:

- Might other cats or animals (indoors or out) be annoying or challenging your cat?

- Have you opened the windows recently after a long winter? Perhaps the sudden rush of odors and other stimuli from outdoors has proved too much for your cat to handle all at once.

- Have you recently moved to new quarters? Is it possible scent marks left by previous owners' cats are getting your cat overexcited?

- Has there been any unusual stress in your household? Be honest with yourself. Your cats may have picked up on a conflict of which you're barely aware, or that you're in denial about.

A Mysterious Case of Aggression: Solved!

Wizard and Angel had always been on friendly terms. So I was puzzled when Wiz started harassing Angel, chasing her into corners, hounding her, and even leaping atop her as she fought to escape. This curious behavior came and went, but it was very intense while it lasted—usually for a few days at a time. What was going on? As the conflict worsened, the stress level in my household rose, and the other cats were constantly on edge, waiting for the next explosion.

My veterinarian was puzzled, too, until one day he called me on a hunch. A trip to the clinic finally revealed the cause: Angel's anal glands were not emptying properly. A cat has two anal glands, or anal sacs, located on either side of the anus. They secrete a smelly fluid that is normally emptied out when the cat defecates. But in some cats, the fluid builds up and thickens, clogging the openings and trapping the secretions inside. The odor of these secretions drives some cats (like Wiz, who seems to be very sensitive to it) to extreme behavior.

The odor of anal gland secretions has intense meanings to cats, with both territorial and sexual connotations. Powerful stuff! Once Angel had her anal glands expressed (emptied) at the clinic, a quick and simple procedure, the problem went away. Now, whenever I see Wiz starting to get too pushy around Angel, I know it's time for Angel to visit the clinic to have her anal glands expressed.

REDIRECTED AGGRESSION

You come home to find your formerly best-buddy cats growling and hissing, crouched in postures of extreme fear and hostility. This can be a real puzzle, and a real heartbreak. It might be redirected aggression—an attack on a substitute for something a cat wants to pummel, but can't get at.

Your cat might have become enraged when he saw a strange cat walking across the yard. Unable to get at his real target, he turned on the closest substitute: the cat buddy lounging next to him. A cat still highly aroused after a recent fight might attack an uninvolved cat who just happens to enter the room.

Humans can be targets of redirected aggression, too. Say your cat's upset or stressed out about something. With no idea how he's feeling, you pick him up for a cuddle, and he lashes out at you and slashes your arm.

Such incidents do not make for family harmony. A serious, sudden, unprovoked attack can cause two formerly friendly cats to become enemies, at least temporarily. An unexpected attack can also cause a rift between you and your cat, threatening your relationship and your bond.

If two cats suddenly seem to hate each other for no apparent reason, and if they are acting aggressive or jumpy, separate them. They need to settle down. Be careful when handling them—you don't want to become the next target. Gradually reintroduce them under pleasant, happy circumstances, perhaps with a meal or treat. Be vigilant for lingering signs of fear or aggression, and be ready to separate them before a battle begins. In severe cases, you may have to reintroduce them just as if one were a newcomer.

Sometimes, a single instance of redirected aggression can have long-term consequences. A cat who's been the target of a vicious, completely unexpected, and unprovoked attack may decide to withdraw, marginalize himself, and avoid contact with any cats thereafter, becoming a social recluse. After any incident of redirected aggression, discreetly observe all your cats, and watch for any unusual behavior or new behavior patterns. Make sure all cats continue to eat, drink, and use their litter boxes normally.

A Case of "Jekyll and Hyde"?

It was a quiet evening. My husband came home from his meeting and joined me and the cats for a chat in the studio. Suddenly there was an explosion of screeching, hissing, and flying fur. I was astonished to see Sir Silver, my most gentlemanly cat, rolling and tumbling with Angel. Silver never fights! The tiff ended almost as quickly as it had begun. At playtime just before bed, neither combatant seemed to hold a grudge, so I chalked up the incident as just another kitty mystery.

The next morning I stepped out into the garage to toss some trash. The place was a shambles—trash cans overturned, trash everywhere, food containers licked clean and scattered. My husband had forgotten to close the garage door and a raccoon had taken quick advantage. Suddenly last night's mystifying catfight made perfect sense. Silver had heard the intruder, tantalizingly close yet maddeningly inaccessible. (The garage is directly beneath my studio.) His predatory and territorial instincts had been highly aroused, but unable to get at his intended prey, Silver took it out on the closest accessible "prey-like" creature—Angel. The kitty mystery was a textbook case of redirected predatory aggression.

Once again, I understood the importance of not meddling in these brief flurries of feline excitement.

DANGEROUS AGGRESSION

Very occasionally a cat will become so aggressive and unpredictable that he becomes a real danger to the people and other animals in his environment. Extreme aggression should never be ignored or tolerated. Schedule a complete veterinary medical work-up without delay. Pain, neurological disease, tumors, and other serious medical conditions can cause a cat to become extremely aggressive, sometimes quite suddenly. Never assume that an unusually aggressive cat is just being mean or bad, especially if the behavior has appeared or worsened suddenly or rapidly. Sadly, sometimes the only humane course for a very ill, very aggressive cat is euthanasia.

AGGRESSION AGAINST PEOPLE

Aggression against people is more common in cats less than two years old: kittens and "teenagers." Like human teens, they're likely pushing the envelope (and pushing your buttons) to see how much nonsense you'll tolerate. Kittens especially consider you one of their littermates and playmates, fair game for kittenish ambushes, pounces, and bites. It's up to you to nip these kittenish challenges in the bud.

Feline aggression toward people—even playful, kittenish aggression—should never be tolerated. In kittens it encourages bad habits, and in adults it's too dangerous. Pound for pound, your cat is one of the strongest animals on earth. Even if he's declawed (de-toed), he's well armed and well equipped enough to do a lot of damage in a microsecond. Children, the elderly, and anyone who can't get out of a cat's way quickly (and cats can move awfully fast) are at risk for serious injury from an aggressive cat.

Pay special attention to body-language clues that signify annoyance or aggression:

- Tail vigorously lashing

- Piloerection—spiky, erect fur

- Low, crouching posture

- Ears and whiskers plastered flat backward against head

- Teeth showing in a snarling expression

- Spitting, exaggerated hissing

- Claws extended or flexing

- Pupils narrowed to slits

- Guttural growls, whines

When confronted with an aggressive cat, even your own, move away calmly and quietly. Don't attempt to approach, pick up, calm down, or otherwise interact with him. Ignore him, and calmly leave the area. If you can figure out what's gotten him so aroused, and you can remove or reduce whatever it is, do so without endangering yourself or closely approaching the peeved cat. Otherwise, just leave him alone to calm down. This can easily take several hours or more.

Outsmarting Skin Scratchers

> "Those who will play with cats must expect to be scratched."
> —*Miguel de Cervantes*

Cervantes overstates his case. It's quite possible to enjoy cuddling and playing with your cat without ever getting scratched. It requires respect, observing your cat's mood and reactions, and stopping before he becomes overstimulated.

Your cat's claws are his first line of defense. With lightning speed, he can extend these supersharp weapons and do a considerable amount of damage. This is a great strategy for a small, solitary predator in the wild, but not in your home. When claws meet human skin, nobody's going to be happy.

In many cases it's the scratchee's own fault. Here's how to get yourself scratched:

- Startle a cat.

- Grab a cat from behind without warning.

- Wake up a sleeping cat by grabbing or picking him up suddenly.

- Shove your hand into a cat's den or hideaway while he's in residence.

- Keep playing with an overstimulated cat, especially using your hands or feet.

- Try to remove a toy (or a captured prey animal) from a cat's mouth.

- Try to force a pill or other medication on a reluctant cat.

- Stroke, pet, or touch a cat's belly after he's given clear signals that he's getting overstimulated (lashing tail, etc.).

- Allow, encourage, or reward rough play, especially involving claws and human hands and other body parts.

- Try to break up a catfight by picking up one of the combatants.

- Handle an injured or very ill cat without donning heavy gloves.

A common misunderstanding leads to many scratches and bites. A cat rolls over on his back, exposing his irresistible, fluffy tummy. He *might* be begging for tummy petting, but he might also just be demonstrating love and trust without really wanting his tummy touched at all. It's not that he doesn't love you; it's just that his tummy is *very* sensitive, and petting or even touching it can overwhelm him. His inner wildcat takes over, and his innate defensive response to touch kicks in. Before you know it, needle-sharp claws are wrapped painfully around your arm.

Outsmarting an Overstimulated Cat

Don't panic, don't yell, and don't get mad. And don't move. Your cat is overstimulated. Any movement, especially trying to snatch your arm away, will make him dig those claws in even deeper. Instead, *very slowly* press your arm toward him. This should momentarily confuse him. (Prey doesn't move toward a predator). Let him calm down until you feel the claws retract. (Try not to squeal in pain.) Disengage your arm s-l-o-w-l-y. It may take a few tries before he's calmed down enough to let you go. Be patient.

If you know *for sure* your cat enjoys tummy rubs, wait until he asks for one by rolling over. But don't start with his tummy. Stroke the back of his head or other neutral spot. Then *slowly* move your hand around, paying close attention to his mood and reactions. A cat on his back is ready for action but also vulnerable, and self-defensive instincts can instantly trump friendship. Keep a close eye on his eyes, whiskers, and ears for clues that he's had enough. Watch for these warning signs:

- Body tenses up.

- Pupils narrow.

- Skin along his back starts to twitch and ripple.

- Tail starts to lash.

- Ears flatten.

- Claws or teeth (or both) sink into your arm. (Oops! Too late.)

Stop at once. Give him some quiet time to settle down.

Outsmarting Ambushes

Stalking and ambushing are natural, normal behaviors, parts of your cat's hunting repertoire. Physically optimized for short bursts of intense speed and exertion, small wildcats need to conserve energy and maximize the effectiveness and efficiency of each hunt. To do this, they rely on ambush, rather than long chases.

Stalking Trade-offs

Researchers at Duke University found that cats, in their stalking motion, have evolved a successful trade-off between energy efficiency and stealth. Most animals optimize their movements to minimize the energy they need to move forward. Not cats—and especially not stalking cats. When a cat slinks along the ground, the actions of his front end cancel out the actions of his back end, which is very energy-inefficient. But the overall result is a much more stable, flowing, stealthy stalking motion. Since cats usually need to sneak up on their prey if they want to eat, this "inefficient" style of movement turns out to be a more successful strategy for them overall.

A hunting cat stakes out the entrance to a prey animal's burrow or den, or positions himself alongside a known prey travel run. Concealing himself as much as possible and moving as little as possible, the cat waits. And waits. A hunting cat has incredible patience and can sit still as a sphinx for what seems like hours. When the prey appears or emerges from its burrow, the waiting cat strikes with lightning speed and overwhelming power. If he's lucky (only two or three hunts out of ten are successful), he eats.

AMBUSHING PEOPLE

Very specific types and patterns of movement are most likely to trigger an ambush:

- Movement across a cat's field of vision

- Movement away from him

- Rapid, jerky movement

- Sudden bursts of movement

- Movement accompanied by interesting rustling, crackling, or squeaking sounds

Knowing this is a great way to improve the fun and appeal of your daily interactive play sessions. The more you know about how prey behaves, and how your cat is wired to hunt, the better you can simulate a real hunt and gratify your cat's predatory inner wildcat.

But when your ankles or shoulders, or the newspaper you're reading, are prey, this feline ambushing strategy can be problematic. For one thing, it's painful. Even in play, claws (even tiny kitten claws) digging into your ankles can evoke gasps and draw blood. Many ambush habits start with mis-training in kittenhood. Many others are caused by simple boredom, or by a cat going through his normal, natural daily predatory cycle.

Here are some ways to outsmart ambushing:

- Never tolerate, allow, or encourage any behavior in a kitten that you won't like in a full-grown cat—especially predatory behavior directed at humans.

- Never let your cat think of toes, fingers, hands, feet, or other body parts as toys.

- Offer opportunities for stalking, pouncing, and ambushing every day, with vigorous interactive play. The fingers you save may be your own.

- If those claws get you, try not to overreact: Don't screech, yell, run after, or punish your cat. Don't let him think it's part of a game. As much as possible, ignore the ambush and quietly leave the area.

- While your kitten or cat is learning that ankles and toes are not prey, wear thick socks or slippers in the house.

If you get ambushed, you probably provoked it, probably inadvertently:

- You (or your toes or fingers) moved or sounded like prey.
- You just happened to be in the right place at the right time, when your cat was in one of his predatory moods.

Observe your cat's daily mood and activity patterns over time. You'll notice that at certain times of the day—around dawn, just before meals, and at his usual playtime—he'll be much more alert, active, and jumpy. These are the times when he's feeling most predatory, and when an ambush is most likely. Watch for:

- Wide-open eyes with large, round pupils
- Tensed body posture, as if ready to spring
- High-pitched, excited vocalizations
- Wiggling rump high in the air
- Tail lashing, or just the tip of the tail lashing
- Positioned for stalk or ambush: half concealed behind a doorway, under a bed (peeking out from the bed skirt, maybe), or up on his climbing tree

Wise owners give the cat tree an unusually wide berth at such times.

AMBUSHING OTHER CATS

As predators-in-training, kittens spend their days stalking, ambushing, and pouncing on their littermates. Compatible groups of neotenized adult cats often continue this play behavior throughout their lives, stalking, rolling, tumbling, and gleefully ambushing one another. It's natural, normal—and usually harmless—play, all in fun. Usually if one cat has had enough, he'll signal his annoyance via vocalizations and body language, and the other cats will wisely back off.

Occasionally a socially clueless cat is slow to get the *knock-it-off* message. Watch the clueless cat discreetly, but avoid meddling unless you're sure that a real animosity is developing or a serious conflict is brewing. Calmly distract one or both cats and divert them to a different, highly pleasurable activity.

RESOURCE GUARDING

Sometimes the fun turns sour and ambushes can cause real rifts among cats. A bully cat might, for reasons of his own, decide to keep all other cats away from a resource he wants to guard, such as a food bowl or the litter box. He'll lie in wait near the resource and ambush any cat who approaches. If he's guarding the only litter box (or group of litter boxes) in the house, you may soon have a stinky problem. (See chapter 5.)

A dominant or bully cat might chase other cats away from food bowls, or intimidate them into waiting until he's eaten all he wants. The victims may not get enough to eat, and Bully risks becoming dangerously overweight. To outsmart the ambushing bully:

- Set up at least two feeding stations, each with food and water bowls. Bully can't guard both at once.

- For between-meal snacking, set up bowls of dry kibble at two or three widely separated locations in your home. Check these often and make sure the bowls are clean and the kibble fresh.

- Before serving meals, escort Bully to a safe hideaway room or other space where he can dine in solitary splendor. Give the other cats plenty of time to eat before letting Bully rejoin the group.

- Try to figure out why Bully needs to guard resources. Is he feeling insecure? Might he be ill? If the behavior has come on suddenly, take him to the veterinarian for a checkup.

Outsmarting Biting Cats

A threatened cat's first choice is nearly always flight. Generally a cat will stay and fight, and bite, only if he sees no chance to flee.

Biting is almost *never* a cat's first choice of offense or defense. An annoyed or angry cat almost always gives a series of signals with his eyes, ears, tail, whiskers, gait and posture, and voice as his annoyance escalates. His first action after this signaling is almost always scratching or slashing with his front claws.

Although there's lots of anecdotal evidence that declawed (de-toed) cats resort much more quickly to biting than fully clawed cats, no hard scientific evidence backs this up. Still, wise owners of declawed cats will keep the anecdotal evidence in mind when approaching their cats in times of fear, stress, pain, anger, or annoyance.

OUTSMARTING PEOPLE-BITING CATS

Cat bite statistics pale in comparison to the whopping number of dog bites. Each year, one million to two million dog bites, mostly to children, are reported to public health authorities nationwide. The number of unreported dog bites is thought to be much higher: Estimates range from four million to five million per year. Half of all kids in the United States will be bitten by a dog before their twelfth birthday.

Cats bite about 400,000 people a year—again, many of them children—and account for about 10 percent of reported animal bites in the United States. The vast difference between dog- and cat-bite numbers has a lot to do with the cat's nature and evolutionary heritage—cats are simply less likely to bite than to flee or scratch. And while some dog owners unwisely encourage aggressive behavior in their pets, it's rare for people to have or keep a truly aggressive cat.

Almost all cat bites occur when the cat is trapped, cornered, startled, or restrained. Trying to hold a panicked, angry, or scared cat against his will is like wrestling barehanded with a chain saw. A frightened cat seeking an escape route is much more likely to think in 3-D than a dog is. He may see the quickest route to freedom and safety as up—and you may be the best climbing structure to get there.

Though more victims go to emergency rooms with dog bites, cat bites are much more likely to become infected. If a cat bite bleeds profusely, chances of infection are greatly reduced, because the bleeding flushes some of the infectious saliva out of the wound. But because of the long, thin structure of the cat's fangs (canine teeth), cat bites are often deep puncture wounds that bleed little, if at all.

According to a review of cat bite cases conducted at the Yale University School of Medicine, 28 percent to 80 percent of cat bites become infected. Any cat bite wound should be considered infected with *Pasteurella multocida,* a bacterium commonly harbored in the feline mouth that can cause severe infection in as little as four to eight hours.

Clean all cat bites immediately with hot water and antibacterial soap. Apply an antibiotic cream or ointment. Seek prompt medical attention for cat bites, especially if the skin is broken. Virtually every cat bite should be treated with antibiotics. Make sure the victim is protected against tetanus by prior vaccination or receives a tetanus shot without delay.

Cat bites pose the highest danger when they occur on the hands and feet, when the victim is immunocompromised, or when the victim has diabetes. If the bite is in a joint, such as a finger knuckle, the injury can be quite serious, with inflammation and bone infection.

If the victim's physician suggests a referral to a hand surgery specialist, don't ignore this advice. Cat bite injuries that involve the tendon sheaths or surrounding compartments in the hand can cause permanent disability. Rarely, infected cat bites cause severe or even fatal complications, such as meningitis, septic arthritis, osteomyelitis, and endocarditis. Cat bites can also cause cat scratch disease (CSD; see chapter 13), although this is rare. CSD usually affects children and immunocompromised persons; in the latter, it can be life threatening.

Cat bites can also transmit rabies. When a bite comes from an NC or OPC, especially a stray or feral cat (any cat with an unknown vaccination status), it's vital to capture the animal for observation, quarantine, and/or testing.

Preventing Cat Bites (and Scratches)

- Avoid contact with unknown cats, especially ferals. Usually they'll avoid you, but they sometimes show fear aggression around people. Any fearful cat is a dangerous cat.

- Avoid confrontations with frightened cats. Unless the cat is in imminent danger, retreat and wait for a better time to pick him up or move him. If you must move him, look for anything thick and heavy (a jacket, a blanket) to wrap around the animal. Don heavy gloves if you can, and keep your hands well away from the cat's face.

- Never try to break up a catfight by picking up one of the cats—even if it's a cat you know well.

- When transporting your cat to the veterinarian, use a secure carrier designed for cats. Avoid cardboard carriers. These are fine for kittens, but an annoyed or frightened adult cat can shred one in seconds. (I've seen it done—*very* impressive.) An inadequately restrained cat might easily get loose, and a dog in the parking lot, traffic noise, or a truck backfiring could cause him to panic and flee. Besides the risk of you and other pursuers being bitten and scratched by the frantic feline, he may run into traffic, or get lost in an unfamiliar area.

- Closely supervise all interactions between young children and cats. Be ready to step in and distract child, cat, or both if either seems to be stressed or behaving inappropriately.

- Never, ever use your hands or feet as cat toys.

- Learn to recognize when a cat has had enough and wants to be left alone. Watch for signals like hissing, a low growl, a lashing tail, or whiskers and ears swept back. Back off immediately.

- Never tease or chase any cat, even one you know well.

- Never clutch or grab any cat, pull his tail, corner him, hold him against his will, or prevent him from escaping a situation he doesn't like.

- Never bother a cat when he's in his "den."

- Never bother or interrupt a cat when he's eating, sleeping, or using his litter box.

- Respect each cat's personal space. Some cats need more space around them to feel comfortable and safe than others do. When in doubt, keep your distance.

- Pet cats only in the direction their fur grows.

- Respect each cat's personal preferences in petting: If he doesn't like having his toes or tummy touched, don't touch them.

OUTSMARTING CAT-BITING CATS

Even in friendly play-fighting, cats can get bitten. Bite wounds inflicted by long, narrow feline fangs tend to be relatively small, deep punctures. On the surface, these wounds close over and begin to heal rapidly. But because a normal feline mouth contains *Pasteurella multocida* and other bacteria, serious infections can fester at the bite site. The result is an *abscess,* a localized collection of pus surrounded by inflamed tissue. Bite abscesses can form large lumps at the bite site. These lumps are usually, but not always, tender and painful to the touch. Cat fur, especially long, fluffy fur, is all too effective at disguising and hiding abscesses. Periodically run your hands gently all over your cats' bodies to check for lumps or bumps.

If your cat ventures outdoors, check more frequently. If you notice anything out of the ordinary, like a large lump that wasn't there the day before, take him to the veterinarian immediately. Infections transmitted by bites can be serious and spread to other areas of the body, including vital organs.

OUTSMARTING PAPER-BITING CATS

Biting and chewing paper, cardboard, and the like is typically a kittenish phase. Most cats outgrow it by the time they're two years old. To outsmart your paper-shredding kitten (or cat):

- Keep objects you don't want shredded out of reach. Stash papers, newspapers, magazines, mail, and the like in drawers or plastic storage tubs with lids.

- Provide acceptable chewies, such as hard kibble. "Dental diet" kibble designed to clean cats' teeth is ideal. The pieces are large and require vigorous chewing.

- Provide small, hard-rubber chew toys (sold for puppies).

- If biting, chewing, and even consuming of nonfood items such as paper, cloth, or cardboard continues or increases, your cat may have *pica,* a tendency or craving for eating nonfood items. See chapter 8 for tips on recognizing and outsmarting pica.

Quick Tips

1. The more cats you have, the more likely you'll have to deal with territorial conflicts, among the most common reasons for aggression, fighting, and stress. No matter how much you love cats, don't go beyond your home's cat carrying capacity.

2. Learn to recognize the difference between a purr and a low growl. The growl is often the first sign of aggression-in-progress.

3. Watch for the "cat stare." An aggressive cat often stares fixedly at his target before mounting an attack. That's the time to distract or divert him, if you can. Be careful; he's already highly aroused.

4. Observe your cats' normal play styles so you can tell vigorous playing from real fighting.

5. Use both your knowledge and your intuition in deciding whether to ignore a tiff or break up what looks like a bona fide fight.

Chapter 8

Food Fussies and Earlobe Chewers

Outsmarting Finickiness

Cats are notorious for "finickiness," a trait made famous by a certain suave, debonair orange tabby spokescat for a major cat food manufacturer. Many owners are secretly proud that their cats turn up their noses at common chow and bargain brands. But even they probably don't want to open seventeen cans until fussy Flossie decides that, tonight anyway, she wants the mystery meat behind door number two.

Winnipesaukee's sensitive tongue immediately knows if her food is fresh, safe, and tasty.

Your cat's priorities in food are:

- *Freshness.* She won't touch even partially spoiled food.

- *Temperature.* She prefers warm food. She can't properly smell food cooler than room temperature (let alone right out of the refrigerator), so she can't tell whether it's fresh. She'll probably reject it. Warm food exudes more tantalizing odors, especially if it's near the temperature of freshly killed prey (95 degrees F), or at least the normal temperature of her tongue (86 degrees F).

- *Texture and consistency.* "Mouth-feel," also influenced by temperature, is an individual preference. Some cats prefer crunchy chunks; other like mush or smooth puree. Some enjoy a variety of textures. Some cats reject all dry food. Some (but not many) reject canned food. Cat food manufacturers

211

spend many research dollars optimizing the shapes, sizes, and textures of cat food to accommodate the maximum number of finicky felines.

- *Taste* (related to a food's animal origin). This, too, is an individual preference. Some cats dislike beef or pork. Most like chicken and adore fish, especially tuna and salmon—the smellier the better. Taste, like texture, is partially dependent on its temperature.

Your cat is keenly aware of salty, sour, and bitter qualities in food, but she has almost no taste buds that detect sweetness. Sugar isn't part of a wildcat's diet. She can detect extremely tiny differences in the kinds and amounts of certain amino acids in her food. This is a survival trait, helping her determine whether food is nutritious, fresh, and safe to eat.

Finicky cats are made, not born. It's fine to serve your cat a few favorite foods most of the time. But to avoid finickiness and to make sure she'll accept different foods should she ever need a prescription diet, offer both canned and dry food, and vary flavors and textures regularly.

Switching your cat's diet too often, too quickly, or too radically can encourage finickiness, too, as well as cause digestive upset. It's false economy to serve your cat "whatever's on sale this week" or to rely on cheap generic or store-brand cat foods. These often contain lower-quality ingredients and fillers that can harm her health over time. Stick to recognized premium brands.

No matter what the advertisements say, your cat doesn't *need* variety in her diet. If you serve her a high-quality, complete, balanced food, appropriate for her life stage and health and approved in feeding trials by the American Association of Feed Control Officers (AAFCO), she'll get all the nutrients she needs for robust health. If she likes a particular brand or flavor, and she's healthy and active, there's no reason not to indulge her most of the time.

GRAZING IN THE GRASS

Unlike dogs, cats don't wolf down the entire contents of their food bowls in two gulps. Cats are grazers by nature. Small wildcats hunt, capture, and consume fresh prey several times a day. It's natural for your cat to eat a little, wander off, and return later for another mouse-sized portion. A bowl of dry kibble for grazing throughout the day is popular with many cats.

Outsmarting Feline Foodies

Your cat's likely going to want a nibble of whatever you're having. She's curious because it smells interesting, and because *you're* eating it. To outsmart your food-begging cat, give in. Usually a lick of butter, a bite of mashed potatoes, or a spoonful of noodles-and-cheese-sauce will satisfy her curious taste buds, and she'll probably wander off to pursue something more interesting. But the more you try to shoo her away from your plate, the more intriguing and irresistible it'll become in her eyes.

Most cats aren't interested in eating more than a token amount of people food. There are exceptions, though. Cats have been known to consume heroic quantities of yogurt, whipped cream, ice cream, cheese (Parmesan seems to be a particular favorite), and even spaghetti. Many cats develop gourmet cravings for such exotic cuisine as cantaloupe, sliced deli ham, chickpeas, or broccoli. Go ahead and give in to her cravings—within reason. Many cats enjoy fresh or lightly steamed vegetables as a complement to their regular diets. Offer fresh chopped lettuce or steamed snow peas, broccoli tops, beans, or carrots. See what your cat likes. But feed her a complete, balanced, cat food, and make sure people food constitutes no more than about 5 percent to 10 percent of her diet.

Fortunately, cats are much more sensible than dogs, who'll eat virtually anything that doesn't eat them first, from garbage to beach towels to bars of soap to tennis balls.

FOOD SNATCHING AND FOOD BEGGING

To outsmart your food-scavenging feline and keep your dinner for yourself:

- Escort your cat to her hideaway room, with her own tasty dinner, before sitting down to your own meal.

- Make your dining room a cat-free zone.

- Let your cat have her fill of sniffing your chow. That may be all she needs to satisfy her curiosity.

- Use a lot of pepper, sharp-tasting sauces, and spicy condiments. Cats hate these.

- Place a few orange slices or half a fresh lemon on your plate. Cats hate the sharp odor of citrus.

Warning: Never let your cat eat any food containing onions or onion pow-der, macadamia nuts, chocolate, cocoa, very salty foods, coffee or coffee beans, or tea. And never, ever let her drink alcoholic beverages, including beer. No, not even a sip.

Keep all meat and poultry bones well out of feline reach. Bones can splinter when chewed, causing serious internal injuries. Dispose of meat and poultry bones in a closed garbage can, outdoors or in a receptacle totally inaccessible to your cat.

- If the food's something that's okay for her to eat, give her a dab of it on a small plate of her own (floor or table—your option).

WATCH THAT TUNA

Tuna enjoys a reputation as the ideal cat food. It isn't, though. Most cats adore tuna. Some even become tuna addicts, refusing other foods in a classic display of finickiness.

An occasional treat of human-grade tuna, or tuna cat foods balanced with a variety of other protein sources, is fine. But too much tuna, or too much in proportion to the rest of the diet, can cause health problems. If all or a large part of your cat's diet is tuna, especially tuna packaged for human consumption, she risks developing vitamin E deficiency that can lead to *steatitis (*yellow fat disease). Symptoms include appetite loss (dangerous for a cat), fever, and hypersensitivity to touch caused by the inflammation and necrosis of her subcutaneous fat.

Because tuna, like most deboned fish, lacks calcium, sodium, iron, copper, and several other nutrients, a cat fed too much tuna can develop serious nutrient deficiencies. There's also a danger that toxic mercury (much more concentrated in tuna than in most other fish) could cause problems over time.

PLEASE DON'T GO VEGGIE

A vegetarian diet *cannot* work for your cat. She's an *obligate* carnivore. She needs meat. She *must* obtain her protein from animal tissue.

For ethical or health reasons, you may have chosen a vegetarian or vegan diet for yourself, getting the protein and other nutrients you need through combina-tions of nonmeat sources. But feeding your cat such a diet imperils her health and life. Cats deprived of meat protein suffer greatly. They often go blind before experiencing fatal heart failure. You can't change your cat's physiology or her nutritional requirements any more than you can teach her not to hunt, or not to breathe.

Refusal to Eat

There are lots of reasons your cat might temporarily refuse food, or sharply cut back on her intake:

- Fever or illness
- Household stress, upset, or noise
- Stuffy nose or upper respiratory problems (lessening ability to smell food)
- Hot or humid weather
- Grief
- Separation anxiety
- A too-rapid diet switch
- Switch to a food she dislikes
- Mouth or tooth pain, periodontal disease (making chewing painful)
- Broken teeth or cavities
- Tongue ulcers
- Sinus problems
- Kidney failure
- Liver disease
- Diabetes
- Inflammatory bowel disease
- Heart disease
- Cancer, including feline leukemia

Skipping a meal or two is usually no big deal for a healthy adult cat. For young kittens, ill cats, and elderly cats, though, missing even one meal can mean trouble. Ill cats and cats recovering from surgery or injuries are also at risk of malnutrition or a depressed immune system if they cut back on the chow. If your cat's started taking just a few bites and wandering off for more than two meals in a row, you may have cause for concern. Call your veterinarian.

If any cat stops eating for more than a day or so, pay attention. Try these special measures to get her back on track:

- Feed her in a separate, private location, especially if you have other cats who may be competing for food.

- Minimize stress and distractions.

- Serve an especially tasty, oily, aromatic canned liver or fish-flavored food.

- Serve cottage cheese or scrambled eggs.

- If she seems reluctant to chew, make her food goopier and more liquid by running it through a blender for a few seconds.

- Heat up her chow. This increases the aroma and may tempt her.

- Sit with her and speak softly and encouragingly. Pet and praise her lavishly when she eats.

- Serve food on a flatter, broader bowl.

- Offer several smaller meals throughout the day.

- Offer tasty treats along with her regular food.

- Hand-feed her.

Don't try to tempt your cat with baby food—a common recommendation because most cats adore the stuff. Many baby foods contain significant amounts of onion powder as a flavoring agent. Onions in any form are toxic to cats.

- Keep mealtimes completely separate from medication and other treatments.

- Don't give your cat any reason to avoid her food bowl. Remove and clean up old or spoiled food promptly.

- Never try to force her to eat.

APPETITE LOSS IS SERIOUS BUSINESS

If your cat is not back to her normal eating patterns after two days, take her to the veterinarian immediately. It's extremely dangerous for your cat to stop eating for that long, or to rapidly and drastically cut down on the quantity of food she eats. Cats can become weak and dehydrated quickly. She may need supplemental feeding and fluid therapy to counter dehydration and to prevent the onset of *hepatic lipidosis* (fatty liver disease), a dangerous accumulation of fat in the liver. Older and overweight cats are particularly susceptible to this often fatal disease.

If your cat has gone more than a few days without eating, your veterinarian may recommend *force-feeding*. This involves injecting, with a syringe-like plunger (there's no needle), a supernutrient paste into your cat's mouth. The formula is so sticky that she can't spit it out, so she has virtually no choice but to swallow it. A very small quantity of such a formula provides the nutritional equivalent of a full meal.

If your cat resists force-feeding and is in imminent danger of developing serious complications, your veterinarian may administer a very low dose of diazepam (Valium) or other benzodiazepine tranquilizer by intravenous injection. Several years ago, a veterinarian discovered, quite by accident, that this drug caused ill cats to resume eating almost like magic. This is considered an emergency procedure, and it's only appropriate when all other attempts to get your cat to resume eating have failed.

Feeding Do's and Don'ts

Do take any loss of appetite in your cat seriously, and monitor it vigilantly. If your cat stops eating for as little as three meals, or cuts back severely on her consumption for more than two days, see your veterinarian at once.

Do serve your cat a balanced, complete, high-quality food made especially for cats.

Do respect your cat's preference for smell, taste, and texture in her food.

Do make mealtimes unhurried, peaceful, and calm.

Do read the labels to make sure your cat is getting the nutrients she needs.

Do be skeptical of label claims and promises.

Who Rules the Roost?

Researchers at the University of Parma in Italy were studying Rome's alley cats to learn about feline social organization. They found that the males are usually dominant—except around food bowls. Males defer to females and even to kittens at mealtime. That's because female alley cats are almost always pregnant or nursing, so they need twice as much food as the toms. The males engage in this unusually chivalrous behavior for a basically selfish reason: Keeping the females and kittens healthy maximizes their reproductive investment.

Do conduct your own research and consult with your veterinarian about your cat's diet.

Do keep up with advances in cat nutrition. New discoveries are being made all the time.

Don't try to make up for a lack of quality time, interactive play, and attention with extra food or treats. Not only will you be contributing to possible obesity, but your cat will see right through your cupboard love.

Don't feed your cat dog food. It lacks amino acids and other nutrients vital to feline health, and has a completely different balance of nutrients.

Don't feed your cat table scraps or tuna packed for human consumption, except as occasional treats. These foods do not provide balanced adequate nutrition for cats.

Don't believe everything you read or hear about "natural" and "organic" diets. Ask your veterinarian for advice.

Don't feed your cat raw meat, especially ground raw meat.

Don't add vitamins, minerals, or other supplements to your cat's diet without consulting your veterinarian.

Don't starve your cat to try to force her to eat something she doesn't like. It's dangerous, especially if she's very young, old, ill, or overweight.

Don't starve your cat in the hope that she'll become a better mouser. It's cruel, unhealthy, dangerous—and it doesn't work. A strong, healthy, well-fed cat is a much more efficient and successful hunter.

Don't give your cat milk or milk products until you're sure she can handle them. Many adult cats are lactose-intolerant; milk gives them diarrhea and digestive upset. But if your cat loves milk and suffers no ill effects from it, go ahead and indulge her. It's a good source of fats and protein.

Water Matters

Your cat's closest ancestors were desert animals. Evolution has endowed domestic cats with the ability to use water very efficiently. Unlike you (and dogs), your cat loses very little water through her skin or in breathing. Instead, her urine becomes more concentrated. She loses almost as much fluid in her saliva while self-grooming as she does through urination.

If your cat eats mostly canned food, she gets most or all the water she needs from her food. (Canned food is about 74 percent water.) A cat who prefers dry food will drink more water. Make sure she always has plenty available.

A ten-pound cat needs about ten ounces of water per day, whether from food or a bowl. Water:

- Helps flush out bacteria and toxins from her bladder

- Helps prevent formation of crystals

- Helps your cat regulate her body temperature, summer and winter

- Is vital for digestion of dry food

- Is vital for proper absorption of vitamins and other nutrients

If your cat's been diagnosed with diabetes or kidney disease, it's especially important for her to drink plenty of water. A cat fountain that offers fresh running water may encourage her to drink more. (See Resources.) These clever electrical devices run water through a small pump and send it splashing into a bowl. The movement of the water keeps it aerated and oxygenated.

The fountain's splashing sound helps remind cats to drink, especially important for ill and elderly cats. Clean the fountain frequently to keep the water healthy and fresh. Before you buy a cat fountain, try to get a look at the instruction manual and see how difficult or easy it will be to take apart and clean. Some fountains (such as the Drinkwell 360) are *much* easier to clean than others. Even perfectly healthy cats often prefer running to still water. Wildcats drink running water (from a stream or creek, for example) if it's available, rather than stagnant water.

Outsmarting Pica (Eating Nonfood Materials)

Your cat licks photographs or leather jackets. She gnaws on wool, wood, or plastic. She eats rubber bands. She sucks on your hair, nibbles on your earlobes, or grooms your skin obsessively.

It's called *pica,* and it may or may not be a problem—depending on the material she chooses, how persistent she is, and your attitude toward her behavior. Some behaviorists think that cats who were weaned (removed from their mothers) too early, or are seeking extra roughage in their diets, are more likely to be affected, but there are no hard scientific data to back this up.

Some pica seems to be breed-linked; Siamese, Himalayan, and Burmese cats are notorious for consuming entire afghans and cashmere sweaters. These wool chewers gnaw and nibble on rugs, upholstery, even the clothes you're wearing. They don't just knead and gnaw, though. They eat the stuff. Fortunately, wool is digestible. But a chewer who prefers synthetic fiber materials risks dangerous internal blockages.

Pica's origins are usually mysterious, but it can result from:

- Hyperthyroidism

- Mouth or tooth pain or infection

- Iron-deficiency anemia

- Nutrient deficiencies

- Bowel diseases that affect the small or large intestines

- Pancreatic disease

- Intestinal parasites

- Stress, anxiety, or boredom

- Desire for attention

- Investigatory or play behavior that gets out of hand

- Obsessive-compulsive disorder

If your cat is a persistent licker, eater, or chewer of nonfood items, and especially if the behavior has worsened suddenly, schedule a veterinarian visit to check for medical causes. But many cases of pica turn out to be similar to what in humans is called *obsessive-compulsive behavior,* caused by stress, anxiety, or boredom.

Cats almost never engage in that canine favorite, coprophagia (eating feces). Most feline pica is harmless, if puzzling. But there are dangers:

- Intestinal blockages that can require surgery

- Damage to your relationship with your cat

- Permanent damage to target objects

Never let your cat chew on, play with, or consume small objects, such as rubber bands, string, dental floss, coins, and small plastic items. Keep them picked up and out of paw reach. If your cat enjoys playing with found toys like milk-jug rings and bread tags, be sure she's not shredding these and eating the pieces. Some cats just bat these around; others can't resist total demolition and consumption.

Never scold or punish your cat for pica. She won't know why you're upset, and it'll likely just confuse and scare her, aggravating the problem. Instead, outsmart her:

- Feed her smaller meals, more frequently.

- Hide or "bury" (under blankets, furniture, or the like) dry kibble or treats around the house so she can forage.

- Offer tiny chunks of frozen fish or a mix of unfamiliar cat treats.

- Hide treats or kibble in treat balls. These small plastic balls (available at pet stores) can be opened and filled with dry food. As your cat bats and plays with the toy, the treats dribble out unpredictably through holes in the ball.

- Hide treats in shoes, or shoe boxes.

- Dab the affected objects with a bitter-tasting substance such as Bitter Apple (available at pet stores and garden centers) or Tabasco sauce.

- Add fiber and roughage to her diet: veggies, bran, psyllium powder, canned pumpkin (plain, no spices), or a high-fiber, crunchy dry food.

On a snowy winter day, Silver, Wizard, and Angel snack on fresh, healthy greens.

- Grow kitty grass. (See the "Kitty Salad" box.)

- Sacrifice an article of clothing or other item in which she's shown a strong interest. If it's not digestible, make sure she's not consuming it.

- Keep wool sweaters, blankets, leather jackets, and other objects of interest picked up and stowed in closets, chests, or lidded plastic storage boxes.

- Drape leather and vinyl furniture and other objects with washable throws or slipcovers.

- Keep potentially harmful objects (rubber bands, small edible toys, photographs, string, yarn, craft supplies) out of her reach.

- If you catch her chewing, sucking, or licking, divert her attention to a play session or interactive toy. Keep a laser pointer (available in pet stores and office supply stores) handy in your pocket for this. Many cats can't resist chasing that bright red dot.

- Consider adopting another cat, or a pair of kittens, to alleviate boredom and give her something else to focus on.

Kitty Salad

Wild and outdoor cats often nibble on grass and other plants. They might be seeking extra fiber, or vitamins and minerals lacking from their diets, plus the antioxidant benefits of chlorophyll. They may be taking advantage of plants' emetic qualities: Cats often vomit after eating plants, ridding their systems of hairballs, worms, and the indigestible parts of insects or other prey. Or maybe they simply crave the taste, texture, and smell of the grass, or a bit of variety. Help make your cat's indoor environment more natural and satisfying by regularly offering her fresh greens.

Every few weeks, start a few pots of wheat (*not* wheatgrass) or oat sprouts, or mixed greens so you'll always have a fresh supply. Pet supply outlets and farm-and-garden stores often carry seed mixes prepared especially for cats. Never use chemically treated seed or ordinary lawn grass seed. The sharp edges of many lawn grasses can injure or irritate your cat's mouth and internal organs. Use sterile potting soil and skip the fertilizer and chemicals. Let the greens grow on a sunny windowsill in a cat-free zone until they're about four to six inches high. Then bring on the salad.

- Train her to do tricks. Many cats enjoy this and respond well to clicker training. (See chapter 12 and Resources.)

- Enhance her environment with extra climbing trees and scratching posts, a (covered) tank of tropical fish, bird feeders (outside a window perch), a window-mounted box she can sit in, or a safe outdoor enclosure with natural features such as cat-safe plants and tree trunks.

- Try cat massage. (Again, see Resources.)

- Play soothing classical or new age music, or a radio tuned to a talk show.

- Offer extra cuddling and petting sessions: ten to fifteen minutes, at least twice a day.

- If the presence of other cats seems to set her off, make sure she has plenty of places to get away and hide, especially up high.

- Reduce noise and stress in your home. This will benefit you, too.

Outsmarting Hair Droolers and Earlobe Chewers

Drooling, kneading, and sucking are deeply instinctual, highly motivated behaviors reminiscent of the close, intimate bonds your cat once shared with MomCat. Extending them to you is a supreme compliment: She's dubbing *you* "Mom."

Some cat owners adore these kittenish behaviors; others find them annoying or disgusting. If you'd rather not have chewed earlobes or drooly hair, gently let your cat know how you feel. Be consistent.

- If your cat chews or licks your hair, tie it in a ponytail or braid (if you can), or wear a hairnet, hat, or scarf with your hair tucked inside.

- If she persistently grooms you or licks your skin (which can be painful), apply a perfume or lotion she dislikes. Or wear long sleeves and socks.

Drooling can be caused by mouth pain, digestive upset, and some neurological conditions. If your cat has suddenly started drooling, or if the scope of the behavior has greatly increased, schedule a veterinarian visit.

- Dab a tiny bit of hot pepper sauce or Bitter Apple on the spots she usually goes for.

- For droolers, wear something that cat slobber won't damage. Or follow the lead of parents of infants: Keep an absorbent cloth diaper handy on your shoulder or lap.

If these tips, tricks, and stratagems don't seem to outsmart your drooler, chewer, licker, or sucker, ask your veterinarian about possible pharmaceutical help. Drugs that have helped cats with pica and other obsessive behaviors include:

- Clomipramine (Anafranil, Clomicalm)

- Fluoxetime (Prozac)

- Diazepam (Valium)

- Amitriptyline (Elavil)

- Buspirone (BuSpar)

These drugs work best in combination with a consistent program of behavior modification.

Quick Feeding Tips

1. Serve your cat her food in unbreakable dishes. Stainless-steel bowls are durable, nonallergenic (some cats are allergic to plastic), and easy to keep clean. Avoid ceramic bowls, especially those made overseas (the glazes sometimes contain toxic materials).

2. Avoid plastic bowls. Aside from the allergy issue, plastic also tends to accumulate pits and scratches that can harbor bacteria.

3. Use wide, shallow bowls to prevent annoying "whisker tickle."

4. For a food gulper who eats too much, too fast, and then vomits, serve meals on a dinner plate. Spread food out in a very thin layer so she can't gulp whole mouthfuls at once.

5. If your cat gulps too much dry food at once and regurgitates it minutes later, put out only a handful of kibble at a time.

6. Or . . . switch to a dry food with larger chunks or different shapes and sizes—anything to slow her down. Kibble in small, smooth, uniform chunks is all too easy for cats to gulp faster than their systems can handle.

7. Or . . . mix some canned food in with the kibble, or add some water to it.

8. Dry food is highly concentrated. Most cats don't need more than about a cup of dry food a day. Read the package label carefully, and make sure your cat isn't getting more calories than she needs. Ask your veterinarian if you aren't sure how to interpret the numbers on the package.

9. Automatic feeders (devices that dispense food into a bowl as it becomes empty) are bad news for cats who tend to overeat. If you have a pudgy puss, dispense measured portions at set times of the day.

10. You'll be tempted—and your cat will try very hard to persuade you—to keep her food bowl constantly full. That's a recipe for a fat cat.

Cat-safe Plants

These plants are "cat-safe" *only* if you grow them yourself, or you're absolutely sure they haven't been treated or sprayed with pesticides, herbicides, or other potentially toxic materials. If you're growing plants of any kind for your cat, use untreated seed and sterilized potting soil only.

Achillea
African violet
Alfalfa sprouts
Alyssum
Basil
Bean sprouts
Buddleia (butterfly bush)
Calendula
Catmint
Catnip
Celosia (cockscomb)
Chamomile
Chervil
Cleome
Coleus
Columbine
Coneflower
Coriander
Cosmos
Cress
Dahlia
Dianthus (pinks)
Dill
Forget-me-not
Heliotrope
Hollyhock
Hyssop
Impatiens
Lavender
Lemon balm
Lemon verbena
Lettuce
Lovage
Marum (cat thyme)

Miniature roses
Mint
Monarda (bee balm)
Nasturtium
Oat grass sprouts
Orchid
Oregano
Pansy
Parsley
Pea (*not* sweet pea, which is poisonous)
Peppermint
Petunia
Phlox
Portulaca
Rosemary
Roses
Sage
Scabiosa
Shasta daisy
Snapdragon
Spearmint
Spider plant
Spinach
Strawflower (helichrysum)
Sunflower
Tarragon
Thyme
Torenia
Verbascum
Violet
Wheat (*not* wheatgrass)
Zinnia

11. If you get a new food bowl, don't just toss the old one and switch overnight. Cats can become quite attached to their bowls. Instead, put food in both bowls, and switch gradually.

12. Dry food can go stale and unpalatable quickly. Close bags tightly, or transfer the food to a storage container with an airtight lid. Use it up before the expiration date. Store in a cool, dry location.

13. Cover and refrigerate unused canned food. Before serving it, though, let it warm up, or warm it briefly in the microwave oven. Don't warm it too much, though; it should not be above your cat's body temperature. Don't keep opened food for more than a day or so.

14. Dogs snacking on your cat's chow? Hang a baby gate high enough for your cat to slip through, but low enough to prevent Fido's entry.

15. Or . . . feed your cat up high, on a countertop or atop the refrigerator. (This also keeps toddlers out of cat bowls.)

Quick Water Tips

1. For water bowl tippers, use a tip-proof puppy bowl.

2. If you have more than one cat, provide several water bowls. Some cats like water chilled; others prefer room temperature. Add ice cubes to some bowls.

3. Cats are sensitive to the size, depth, and width of their water bowls. If your cat refuses to drink, serve water in a selection of bowls, and stick with the one she chooses.

4. Many cats prefer to drink water away from their food. Use separate food and water bowls (not a connected set). If your cat isn't drinking enough water, place a water bowl several feet from her food bowl, or near a favorite napping spot.

5. Ceramic bowls keep water cool longer—great for hot weather.

6. Cats are extremely sensitive to the smell of chlorine. If your water supply is chlorinated, let tap water stand in an open bowl for twenty-four hours before serving it.

7. If your cat turns up her nose at water, serve bottled water.

8. Or . . . add extra water to her food.

9. Or . . . flavor her water with a dab of tuna oil (from canned tuna).

10. Every day, empty each water bowl, rinse with hot water, and wipe completely. Rinse, rinse, rinse. Detergent or soap residue can offend your cat's sensitive taste and smell, and may keep her from drinking the water she needs. If you wouldn't drink the water in your cat's bowl, why should she?

Quick Treat Tips

1. Keep treats under control. It's all too easy to dispense more than you realize.

2. Instead of cat treats, buy a different brand of dry food from your cat's usual fare and use the individual kibbles as treats. It's much cheaper, and your cat won't know the difference.

3. Or . . . use dental diet kibbles (available from your veterinarian, or use a commercial dental diet) as treats. Your cat will be cleaning her teeth while she thinks she's just having fun.

4. Make your cat work for every treat. Toss treats high in the air, down from the top of the staircase, or across the room. Some talented cats catch tossed treats in midair. Make it a fun game—but leave her wanting more.

5. Hide treats around the house for your cat to discover while you're away. Put them in unlikely places: at the top of her cat tree, under the sofa, behind a chair, on a windowsill, high atop a bookcase. It'll help stave off boredom as well as let her exercise her foraging skills. (To preserve freshness and avoid encouraging rodents, remember your hiding places and make sure your cat finds them.)

6. Mix some dry kibble treats in among some smooth rocks and pebbles in a bowl or on a large tray. Your cat will enjoy the challenge of ferreting them out.

Chapter 9

Psyching Out Fluffy: Feline Emotions and Personality Quirks

"In my house lives a cat who is a curmudgeon and cantankerous, a cat who is charming and convivial, and a cat who is combative and commendable. And yet I have but one cat."

—Dave Edwards

Understanding and Accepting Your Cat's Unique Personality

Just like our friends, spouses, and in-laws, our cats come in complex, occasionally mysterious packages, complete with individual differences and psychological quirks. Your cat's personality and temperament depend on some combination of his early experiences (nurture), especially between three and fourteen weeks, and his genetic heritage (nature). There are difficult cats, oddball cats, quirky cats, clingy cats, bouncing-off-the-walls cats, therapist cats, cats in dire need of therapy, hard-to-read cats, nervous, twitchy cats, and cats possessed of Zen-like calm.

Some feline personality types are easy to live with; others call for extra measures of understanding, patience, tolerance, acceptance, and love. Instead of focusing on the downside of your cat's quirks, glory in the ways he's unique.

Don't expect your cat to be like any other cat: not the pedigreed marvel down the block, not the perfect, precious pet of your childhood. He's his own cat, and that's that! Get to know him for who he is. When you first meet your cat and begin to forge a mutual bond, he'll start sending messages addressed to you and you alone. Pay attention. The better, the more accurately, the more sensitively you read your cat, the more successful and satisfying a relationship you'll share.

Each cat has his own personal taste in companionship (feline and human), desire for physical and emotional closeness, love of (or avoidance of) solitude, napping spots, play, and petting. Your cat might be a one-person animal who'll never warm up to anyone but you. Or he might be everyone's best buddy. There are cats who enjoy the social whirl, and Garboesque felines who "just vant to be alone." As much as possible, observe, understand, and respect your cat's social preferences.

And don't think you can change him. Though there are always exceptions (cats being cats), you can't convert your one-person model to an *I-love-everybody* cat. If you try, you'll likely get a confused, unhappy, resentful cat. Is your cat the shy, retiring type who retreats under the bed whenever visitors arrive? The worst thing you can do is scoop him out and shove him in your guests' faces. It's rude and inconsiderate. *You* wouldn't like it, would you?

Depending on his early experiences, your cat might greatly prefer either men or women. Some cats even have a decided dislike for one gender, probably because of an unpleasant early experience. If your cat is one of these, his prejudice can be overcome, or at least modified, with patience, persistence, and plenty of gentle, nonthreatening interaction with an understanding representative of the suspect gender, liberally garnished with tasty treats and lavish praise. Never underestimate the power of telling your cat—over and over and over—how beautiful and special he is.

SEND IN THE CLONES?

If you have more than one cat—even of the same breed, even littermates—don't expect them to share looks, temperaments, or personalities. They might. But then again, they might not. Even the most closely related cats are far from clones. In fact, even clones aren't clones.

CC ("CopyCat"), the world's first cloned domestic cat, had already displayed, at just over a year old, looks and personality much different from Rainbow, the cat from whom she was cloned. Chunky Rainbow is a brown mackerel-patched tabby-and-white shorthair, mostly white with striped patches of black, brown, tan, and gold. Sleek CC is also mostly white, but with a completely different pattern of grayish stripes. Rainbow is mellow, quiet, and reserved; CC is playful, curious, and outgoing. Clones, anyone?

UNDERSTANDING FELINE SOCIAL LIFE

Cat society is not hierarchical in the sense that the military, or dogs, would understand. It's a dynamic, fluid, ever-shifting arena of space and time sharing; tiny, graceful adjustments; constant social signaling; and scent mingling through *allo-*

Tabigail grooms Silver, sharing their familiar, comforting group scent. Later, Silver will return the favor. This mutual grooming is called allogrooming.

grooming (mutual grooming) and touch that define and reinforce the complex web of interrelationships.

There are plenty of status clues for those in the know. For example, dominant cats tend to lounge in the middle of rooms, claim the top spots on climbing trees, and hog the laps of favored humans. But cat societies generally lack a true "alpha cat," a "maximum leader" to whom all the other cats defer at all times. Neither you nor any other human can become the alpha in cat society. As a human companion to cats, you may be welcomed into their social circle as an honorary cat, but you're not going to be in charge—and the sooner you realize it, the better.

Still, social status and social roles are very important to cats. Studies of cat colonies in dockyards and farmyards show that cats have organized social structures and can live harmoniously in fairly large groups, as long as there's a dependable food source and plenty of resources to go around. Cat societies are much more focused on territories, and the resources they offer, than on the status of individual cats. If there's plenty of everything, individual social status becomes relatively unimportant.

Dominique, Princess Dagny, Bunny, and Petunia have created a complex, dynamic, and mostly peaceful social order.

If resources are scarce, social position becomes much more important. Stronger and more aggressive cats guard and monopolize food and females, drive off younger and weaker cats, and fight fiercely to keep potential competitors at bay. The fewer resources available, the pricklier cats become, and the less inclined they are to get along harmoniously.

In response to variables in their environments and the mix of temperaments in the group, cats form a variety of social structures, just like we do:

- In a *despotic hierarchy,* there's one top cat, and the rest are level-two cats. There might also be a pariah cat, picked on by everybody. Not all groups have a pariah. Behaviorists have mixed opinions as to whether "pariahs" actually exist, or are just human misperceptions of more complicated relationships.

- In an *oligarchic hierarchy,* a few cats rule over the rest, usually benignly.

- Many cat groups are *matriarchal.* The top cat is a senior female, either the mother of some or all of the others, or occupying the mother role. Usually she's a tough but benevolent despot, a kind of Mother Superior who stays above the fray and makes sure her charges are happy, healthy, respectful, and properly groomed at all times.

- In a group with intact (unaltered) males and females, the biggest, boldest males will grab and hog as much of the prime territory as they can defend from other males. But that won't necessarily make them top cats. Likely, a tough, fearless female will still rule the roost.

- Intact male cats are notoriously prickly and disinclined to be friends with one another. But once neutered and freed from the pressures of competing for mates, groups of males can form almost kitten-like bonds, rolling and tumbling happily, grooming each other, and sleeping curled up together in the sun: male bonding.

- In some groups, it's difficult to determine who the leader is, if indeed there is one. They live together harmoniously, peacefully sharing space and resources.

- In some groups, the leadership roles change, based on the situation.

- Groups in which cats constantly come and go display ever-shifting structures, with a core group of a few cats (generally senior females) managing the colony's continuity.

Within their social groups cats, like people, have confidants and buddies, friends and enemies. To conduct everyday affairs, they use a rich set of signals: body language, touch, visual cues, scent, and sound. Despite myths of feline aloofness and unfriendliness, cats are a high-contact species. Among themselves, cats living together frequently touch noses in greeting, rub cheeks and foreheads, sniff one another, brush gently while passing side by side, and engage in fond, intense allogrooming. These tiny, graceful rituals are repeated dozens of times a day, and all assist in the mixing, mingling, and refreshing of the familiar group scent.

MomCat—or a senior female who's assumed that role—will often physically hold down even grown cats among her charges for a full daily bath. Wise kitties accede to MomCat's wishes—and have the soft, shiny coats to prove it. In homes with groups of compatible, friendly cats, a patch of warm sunshine will often be graced by two or more busily allogrooming cats. Does your cat extend these little grooming rituals to you? Congratulations. You're an honorary cat.

QUIRKS AS WARNINGS

Glorying in your cat's unique personality doesn't mean letting him walk all over you, though. Stealth claw attacks, urinating on the sofa, serious aggression, biting, antisocial behavior—that's not a cat who's just being his cute, quirky self. He may be extremely unhappy, in pain, or seriously ill. Schedule a visit to the veterinarian immediately, especially if the misbehavior appeared suddenly or is uncharacteristic. Aggression and attacks can result from poisoning, neurological disorders, or other rapidly worsening diseases.

Many types of feline misbehavior, and oddball behavior, are designed to get your attention. Your cat is trying to tell you something and you've ignored it. For example, a cat with chronic or recurring health problems (such as cystitis or urinary tract issues) will often learn to overtly "misbehave" right in front of you when he's feeling bad or having a flare-up—just to make sure that he gets your attention. Never punish him for this! Be grateful that your cat has learned to communicate his needs to you. Listen, observe, learn—and then take appropriate steps to fix his problem.

Outsmarting the Clingy, Needy Cat

> "Cats are our last best chance to have a dysfunctional relationship."
> —*John Bush*

MY CAT KNEADS ME! reads a popular T-shirt. There's nothing quite so satisfying and comforting as being kneaded and needed by this most independent of creatures. An extremely close bond between a cat and his special person is considered sweet and touching, certainly not a danger sign. Owners brag about "Velcro kitties" who can't bear to be apart from them. A clingy lap cat can be a godsend for a lonesome disabled or housebound person. Telecommuters gladly share their home offices with cats who supervise every keystroke and bask in the monitor's warmth. A clingy, cuddly one-person cat offers his special person irreplaceable gratification and love, inspiration and company, warmth and an always-ready conversation buddy. As long as life proceeds normally, it's not a problem.

But life has a way of throwing curves. What if you get hit by a bus on the way to the supermarket? Or have to spend six weeks in a hospital? Get divorced and have to move out—without your cat? What if you bring home a new baby who requires some of the care and cuddling your cat is accustomed to? What if your cat becomes ill and has to spend a week at the veterinary clinic? That sweet, cuddly, closely bonded cat is suddenly a needy, confused, deeply unhappy animal. He's devastated, inconsolable. He's likely to:

- Become depressed, weakening his immune system, risking infection and illness

- Stop eating, risking his health and even life

- Engage in nervous, destructive self-grooming or self-mutilation

- Hide, withdraw, refuse to interact with anyone

- Refuse to use his litter box

Those entrusted with his care may find his clinginess and neediness, and the misbehavior caused by the separation crisis, puzzling, annoying, and off-putting.

For an overdependent cat, even a brief interruption or alteration in his relationship with his person can be highly stressful. His eagerness to bond may mean he's probably less emotionally resilient than other cats, and more susceptible to stress.

As much as you enjoy sharing everything with your One Special Cat, it's only fair to gently help prepare him for possible separations. You can't change his personality or make him something he's not. But it's smart, and kind, to gradually lessen his overdependence and attachment. Here are some ways to help ensure that your clingy cat will have something to fall back on should a separation be necessary:

- Consider a companion cat or a pair of kittens. This will help give your cat something to focus on besides you.

- As much as possible, gently include him in your activities with other people—especially those who would assume his care if something happened to you.

- Increase daily interactive play to burn off excess energy and anxiety. If at all possible, have another person conduct some of the play sessions.

- Let him spend nights in his safe hideaway. Make the hideaway comfortable and luxurious. Play soft, soothing music such as harp, light classical, or peaceful "new age" sounds.

SEPARATION ANXIETY

Separation anxiety is a more common problem in dogs than cats. But a needy, clingy, overdependent cat can show some of the same signs: twitchy uneasiness to outright panic, destructive behavior, desperate vocalizing, and house soiling when he's home alone. Because anxious cats tend to withdraw and become depressed, while dogs rip the place to shreds, cat owners often miss signs of separation anxiety.

Many veterinarians and behaviorists have had success using the drug clomipramine (Clomicalm), together with a program of behavior modification, in cats with severe separation anxiety. The usual method is to start the drug along with the behavioral modification, and taper it off once the cat shows significant improvement. Whether or not your veterinarian recommends using this or another drug, there are some simple ways you can help outsmart your anxious cat:

- Don't make a big production out of going out.

- Cut down on the number of clues before you leave the house.

- Leave your keys or purse near the door so you can grab them quickly and quietly.

- Pocket your keys an hour before you plan to leave.

- Change your coming-and-going routines.

- Put your jacket or coat on after you get out the door.

- Don't say elaborate good-byes.

- If you have a leave-taking ritual, drop or change it—anything to break the pattern of your cat becoming distressed at your leaving.

- When you come home, don't make a big deal out of it.

- When you get back, your cat will probably be excited and happy. But wait until he's calmed down and is behaving normally before offering praise or a treat. Reward his normal behavior, not his neediness.

- Even if your cat did something you'd rather he hadn't while you were away, don't make a big deal out of it. (And remember, punishment never works.) Your anger will just confuse him and increase stress, making everything worse.

- To better accustom your cat to your coming and going, take several short fake outings. Get your purse or keys, go out quietly, and return a minute, or five minutes, or fifteen minutes, later. Gradually lengthen your absences so your cat will discover that nothing really terrible happens when you're gone. Vary the schedule and length of time you stay out.

- When you need to be away, escort your cat to his hideaway for the duration. Make sure you've already established his comfort and pleasure in spending time there alone well ahead of time.

- If your cat needs to spend time away overnight or longer, send along an article of your unlaundered clothing and a cozy, cave-like cat bed to put in his cage. If medically appropriate, send along a supply of his favorite food and treats and instructions for how he likes them served. Make sure the boarding person, veterinarian, and staff know how important this is to you. When a clingy, overdependent cat is ill or injured, it's especially important that his emotional needs are tended as closely as his physical needs.

Outsmarting the Stressed-out Cat

"Some cat owners have what I call the St. Francis syndrome: 'Anyone can live with a perfectly behaved cat, but it takes a real saint to put up with mine.' If the behavior problem is solved, they lose their sainthood."

—*Myrna Milani, DVM*

How your cat handles the inevitable stresses of life depends a lot on his personality, basic temperament, age, and experiences. Just as in humans, a shy, introverted cat is likely to handle stress by hiding, or becoming fearful or obsessive, while an extroverted, outgoing cat might become destructive or aggressive under the same circumstances.

Knowing your cat and constant observation are key. Has your friendly, extroverted cat suddenly become nervous and twitchy? Has your introverted cat suddenly withdrawn even farther into himself, sitting immobile as a sphinx for hours? It might be stress. The most common sources of stress in cats are:

- Other cats—especially newcomers and other intruders in his territory

- Moving to a new home—it's like you moving to another planet

- Losing a beloved owner or having to move in with a new family

- Illness

A reasonable amount of caution, especially in unfamiliar situations, is a normal, healthy reaction. Healthy, sensible fear has allowed cats to survive as long as they have. When your cat shows signs of fear or stress, look around and see if he has a good reason (to a cat's mind, if not to yours) for his reaction. Chances are, he does.

A stressed-out cat might show his unhappiness in a number of ways:

- House soiling, outside-the-box behavior

- Spraying, marking

- Excessive self-grooming and self-mutilation

- Immobility (depression)

- Hiding

- Aggression, toward people or other pets

- Wailing, crying, howling, other excessive or continuous vocalizing

- Hypersalivation, drooling

- Refusal to eat; picking at food, finickiness

- Pacing, restlessness

These signs can also indicate illness. Schedule a visit to the veterinarian without delay. Removing the stressor and providing your cat with a safe hideaway, extra attention, and daily vigorous playtime can all help reduce stress. In serious or persistent cases, a course of a calming medication can help, if used in conjunction with behavior modification, stress reduction, and desensitization. Ask your veterinarian for advice.

When she's feeling stressed out, Dandelion retreats to her "safe space" under the bed.

Outsmarting the Scaredy-cat

Scaredy-cats are often highly sensitive. They're easily startled, nervous, twitchy, and quick to hiss or paw-swipe at the least provocation. They react edgily or fearfully even after the initial trigger is long gone. This can keep other cats on edge, too. That's why it's so important to avoid reinforcing and rewarding their fearful behavior.

What are some of the *worst* things you can do with a fearful cat?

- Hold and cuddle him when he's most fearful.

- Try to console him, speaking in soothing tones.

- Carry him to near the object of his fear: "Look, Kitty. There's nothing to be afraid of here."

- In other words, make a big deal about the fear, the stimulus for the fear, and the cat's fearful reactions.

While they might seem kind, empathetic, and natural, these reactions reinforce your cat's fear and his fearful behavior pattern by telling him that you agree that there is something to be afraid of.

What's the *best* thing you can do? Act casual, nonchalant, unconcerned. Go about your ordinary business. Show your cat by example that there's nothing at all to fear.

THUNDERPHOBIA

This is more commonly a dog problem. But a few cats become extraordinarily fearful at the first rumble of thunder. And while dogs often go manic and tear up the house, scared cats are more likely to seek deep cover, hide, and tremble quietly. But they're still suffering, and some behaviorists think it's worthwhile to try to desensitize fearful cats to the stress and racket of thunder, lightning, hail, and wind. Some recommend exposing the cat to increasingly frequent and loud playing of audiotapes of thunderstorm noises. But since thunderstorm fears can be triggered by any one, or a combination, of so many stimuli other than noise (including vibrations, barometric pressure changes, flashes of light, odors, and more), this may not be fully effective in desensitizing your thunderphobic kitty.

Silver is alarmed at the first rumblings of thunder.

A truly tough case might require behavior modification, desensitization, and the use of calming medications such as alprazolam (Xanax). A course of buspirone (BuSpar) may help your cat settle down enough that you can conduct a successful behavior modification program.

What shouldn't you do? Don't encourage your cat's scaredy-cat behavior. As much as possible, stay calm and ignore the thunderous racket. ("Calm owner, calm cat.") Letting him jump in your lap, cowering and whimpering, while you console him could reinforce his fear—you're rewarding him with attention for being scared. Let him hide if he feels he needs to. He'll come out when things quiet down.

OTHER WEIRD PHOBIAS

Cats have shown extreme fear of a variety of things: clothing, ceiling fans, toys, machines, certain people, and more. To outsmart a phobic cat:

1. Remove the offending object from the cat's world.

2. If that's impossible, try a slow, gradual program of desensitization. Act casual, nonchalant, and calm. Your cat will pick up on your unconcern.

3. Conduct fun play sessions or treat times in the room or area where the object of the fear is. Associate that spot with fun and good things.

Sometimes it's a good idea for your cat to be afraid of something. Though lots of cat owners complain that their cats are terrified of the vacuum cleaner monster, is that so bad? Do you want your cat underfoot, chasing the vacuum cleaner while you're using it? Same goes for power tools, noisy machinery of all kinds, the furnace, clothes dryer, washing machine, and kitchen stove.

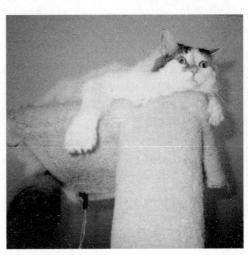

Angel is afraid of . . . something!

DESENSITIZATION

If you know what your scaredy-cat's afraid of, and you can control the level of it he's exposed to, you can usually, with patience and sensitivity, help him to be less fearful. Depending upon the cat, and what he's afraid of, though, you may never get him to completely ignore the object of his fears. Don't push him.

Desensitizing works the same in cats as it does in people. A very slow, gradual, patient approach is vital.

1. Expose your cat to a very low dose of whatever it is that frightens him— so low, in fact, that at first it doesn't provoke his fear at all.

2. Very gradually, increase his exposure.

3. Accompany the exposure to the feared object with happy, pleasant experiences—tasty treats, play, cuddling.

4. If your cat has an overly fearful episode, back off a step and reduce the exposure level.

CAGE DESENSITIZATION

To a formerly outdoor or feral cat, the ordinary sights, sounds, and odors of your household can be very scary and disorienting at first. Cage desensitization can help acclimatize him to life indoors. It's similar to cage confinement described in chapter 5 for attitude adjustment of out-of-the-box thinkers.

Use a cage large enough for a litter box, cozy, cave-style bed for hiding, and food and water bowls. For an extremely fearful feral cat, start with the cage covered, in a quiet part of the house. Play a radio softly. Observe the cat closely, and speak softly to him as often as possible. Spend time just being with him. Read aloud to accustom him to the sound of your voice. As he allows, let him out of the cage for handling, petting, and interaction. Keep the door to the room closed while he's out of the cage, so you have full control over what he's exposed to.

Gradually, as he starts to relax and become more comfortable in your presence, move the cage to a busier part of the house. Be sure he has a place to hide in his cage. Let other cats and pets approach politely, but let the caged cat set the pace. As he becomes comfortable, let him out of the cage in one room initially, and then gradually expand his territory.

Outsmarting the Poorly Socialized Cat

Bunny is a social disaster. Adopted from a shelter at the age of two years, she's a Velcro kitty—clingy, snuggly, and affectionate with humans, especially her special person. But among cats, Bunny's clueless. She doesn't speak or read cat. She misinterprets common cat body language and communication signals. She sees friendly overtures as threats, aggression as desire to play, *leave-me-alone* cues as invitations to a fight.

Bunny never, ever misses a fight. The slightest hint of a snarl or wail has her leaping heedlessly from her owner's lap, eager to join the action. No matter who's fighting (or just playing around), Bunny always jumps right in, paws slashing. She always gets the worst of it. Her sweet pink nose is usually marred with at least one scratch from her latest altercation.

Cat-to-cat social skills, like people skills, are learned, not completely instinctual. Bunny is a sweet, loving cat, highly socialized to people, but a feline social disaster. If you adopt a kitten who's going to share a home with other cats, make sure he spends as much time as possible with his mother and littermates. If possible, he should remain with his original feline family for at least twelve weeks. During this time, MomCat conducts daily lessons in proper cat behavior, communication protocols, and hunting skills. She also passes along much other feline wisdom, including tips on social skills.

Kittens who leave their family circles too early miss an irreplaceable opportunity to develop and refine their cat-to-cat communication skills and to learn a multitude of behavior adaptations and subtle feline communication strategies. As the litter plays and pounces together, the kittens are hard at work, learning:

- How feline hierarchies are structured

- How to rank their siblings

- How to recognize the signs of rank within the hierarchy

- How to deal with older cats and higher-ranking cats

- How to show deference and respect while still maintaining face

- How to tell an invitation to play from a dare to fight

- . . . and much, much more.

Bunny's early years are a mystery. But it's a good bet that she was taken much too early from her momcat and littermates.

SOCIALIZATION HAS ITS LIMITS

A cat who stalks or kills a small or pocket pet (also known as prey animals: birds, rodents, rabbits, guinea pigs, snakes, reptiles, or hamsters) is not poorly socialized. He's just being a cat. No amount of exposure, socialization, or training will enable your cat to tell the difference between prey animals he's "allowed" to kill and those he's supposed to befriend.

If your family includes such pets, use some common sense and skip the socialization. Your cat's been honed by millions of years of evolution to chase, stalk, and kill them. The predatory instincts and behavior patterns hardwired into every cat's brain are powerful and compelling. Chasing, stalking, and pouncing are completely instinctual. No amount of socialization can exterminate them.

If you choose to live with both, keep prey and predators completely separate, with a closed door (at the very least) between them. For the sake of the mental health of your hamsters and other prey-animal pets, don't let them even smell your cat. Perhaps your cat has no realistic chance of getting at the small critter, but the prey animal doesn't know that. All she knows is that there's a fearsome predator close by. It's not fair or healthy to subject her to the constant stress.

Outsmarting the Jealous Cat

> "A relationship with a pet is, for many people, a sort of sacred totem that they're reluctant to change. Some owners restructure their lives and homes around the behavior of a 'jealous' cat because it serves their own emotional and psychological needs. Some owners with low confidence see their cat's 'jealousy' as an affirmation of their own worth."
>
> —*Myrna Milani, DVM*

The green-eyed monster usually appears in highly possessive cats, and in those closely bonded to a special person. Possessive, bossy, pushy, and "only-cat" cats are most likely to be peeved when they find their exclusive use of treasured resources (your entire home, your exclusive attention) threatened by an interloper: new baby, spouse, boyfriend, another cat or pet.

Any cat can get jealous, though. Some behaviorists feel that jealousy is a territorial issue, and that a jealous cat sees preferred humans as parts of his territory: possessions to be defended from competition like any other valuable resource. A jealous cat can make you feel mighty important, treasured, and appreciated. But he can also make your life miserable.

Feline jealousy can focus on objects, space, people, and personal attention. Say you have a cat who's closely bonded to you and is used to having his way. As long as you devote all your attention to him, he's fine. But if he catches you spending too much time working on the computer, talking to another person (even on the phone), or smooching somebody on the sofa (horrors!), you risk the consequences. He'll try to split up the conversation, distract you, or separate you from your honey, and demand attention as if you are a mere chattel—a valuable object that he feels entitled to claim, possess, and dominate. He may even urinate on your computer keyboard. He has no intention of letting you become involved in *any* competing relationship. The very idea!

It's a common situation: A young lady, living alone, has lovingly hand-raised her dear cat, who's now closely bonded to her to the point of overattachment. Dear Cat feels that he owns Young Lady, and naturally expects her to honor and obey him and gratify his every whim. With a male cat and a female owner, the relationship can get quite . . . complex. DC and YL aren't just friends, but mates, and YL is also MomCat. DC sucks on YL's fingers and earlobes, grooms her hair. It's a very intimate relationship, and all is well . . .

. . . until Young Lady meets Mr. Right.

The flurry of sulking, litter box misses, shoe marking, leather-jacket spraying, and general feline outrage that follows needs no elaboration.

But look at it from Dear Cat's point of view. Jealous or possessive behavior is often a cat's desperate attempt to communicate that he's under stress. You love your new boyfriend—but you owe your unhappy cat something, too. Anything you do to defuse and calm the situation, increase your cat's confidence level, restore his faith in the integrity of his territory (including you), and minimize his stress will help.

Because a jealous cat probably feels compelled to defend your entire home from whatever he thinks is a threat (your boyfriend, for example), one of the smartest things you can do is make his job easier. While he gets used to the new circumstances, make his hideaway his core territory. Escort him there at bedtime, and any other time he's feeling anxious. Make the hideaway the site of treats, play, and special one-on-one bonding times.

If the jealousy trigger is a new person in your life, introduce NP and your cat gradually. Let your cat make all the moves. (He was there first, after all.) Don't rush him, or force him to be friends before he's ready. Keep initial meetings brief and pleasant, and then escort your cat to his hideaway.

We usually have fair warning when a new person or pet will be moving in. Some sensitive, cat-focused planning and preparation can greatly minimize the shock that can send an unsuspecting cat into fits of jealous rage, misery, and misbehavior.

NEW BABY ON THE WAY?

As soon as you know baby's on the way, start preparing your cat for the happy addition to both your lives.

- Let him see, smell, and touch baby clothing and supplies and nursery furniture so these will be familiar by the time baby arrives.

- Use baby lotion and powder on yourself. Give your cat lots of treats, cuddling, and attention as he samples these new scents.

- Let him visit the nursery as you're preparing it. Leave the door open; the room will soon lose its mystery (and thus its appeal). Install a screen door for when baby is in residence. Through the screen door, your cat will be able to easily watch and smell what's going on (reducing the tantalizing, irresistible mystery of The Closed Door), but won't be able to interact unsupervised with baby.

- Pamper your cat with a special new deluxe climbing tree or cozy fleece bed. Make sure his home is filled with attractive opportunities for play, lounging, and exploration.

- Don't change your cat's everyday routines any more than absolutely necessary. Keep his world as stable and familiar as you can.

- Keep a cheerful, upbeat attitude. Let your cat pick up on your joy at the approaching happy event. Talk about "his baby."

- If you can get one (maybe from a friend), play a recording of typical baby sounds for a few minutes every day so your cat can get used to them.

- Let your cat play with some cat-safe baby toys (squeaky toys, rattles). Put them in his toy box for him to discover.

- Ask a friend who has a young infant to visit so your cat can actually see a baby. Let your cat watch (he'll probably be *very* interested!) as you hold, cuddle, and talk to the baby. At the same time, your spouse or other familiar family member should snuggle and play with your cat, offering tasty treats and making the baby's visit a pleasant, happy experience. Then change places. Fuss over your cat while other family members admire your friend's baby.

- As soon as your baby is born, send home an article of clothing or a receiving blanket carrying the baby's scent. Let your cat sleep on or with it.

- Once baby is home, don't leave her unsupervised with your cat, no matter how gentle and reliable your cat may be. Include your cat as much as possible in your baby care activities, but use common sense, and keep that screen door closed. Remember, cats *react* instantly to stimuli. Although a cat snuggled up to a sleeping infant makes a charming image, if a sudden loud noise, bright light, or other stimulus startles the cat, the baby could be badly injured by the cat's instinctual defensive reaction.

NEW PERSON (NP) MOVING IN?

- Associate NP's visits with pleasant, happy experiences for your cat.

- Include your cat in activities and conversation while NP is around. Watch TV together. Watch cat videos together. Have NP sit on the floor and blow bubbles.

- Scent mingling and scent familiarization are key to a happy relationship. Before NP visits, if possible, obtain an article of clothing carrying his scent (an unlaundered towel, sweatshirt, or T-shirt is great). Let your cat sleep or lounge on it.

- Keep a container of tasty treats near the door. Whenever NP visits, have him dispense treats.

- If NP is unfamiliar with cat etiquette, teach him how to politely approach an unfamiliar cat: Crouch down or sit on the floor, extend a hand, and let the cat make the next move. Advise NP to avoid staring directly at the cat. (It's considered rude, and a bit threatening.)

- Have NP visit at dinnertime and feed your cat. The first few NP-served meals should consist of extra-tasty foods.

- Teach NP how to use your cat's favorite interactive toy. Have NP conduct a brief play session as soon as he arrives, and again before he leaves each visit.

- Keep a positive attitude. Your cat will pick up on your positive feelings and eventually transfer them to NP.

Outsmarting the I-Hate-All-Other-Cats Cat

Some cats should be "only cats." Because of quirks in their personalities, bad experiences during their upbringing, or lack of socialization with members of their own species, they simply despise all other cats. If you have such a cat, respect his preferences. Although it's theoretically possible to gradually bring him around so that he can cohabitate with another feline without committing kitty murder, it really isn't fair. You may have to accept that for your cat's lifetime, your home's cat carrying capacity is precisely one. Or consider giving your *I-hate-all-other-cats* cat a space of his own within your home, where he can enjoy his preferred lifestyle free of other feline influences. You can always share your extra love with homeless cats awaiting adoption at a local shelter. Shelters are always looking for volunteers to gentle, cuddle, care for, and socialize their charges.

You may be tempted to introduce a kitten. Your *I-hate-all-other-cats* cat may surprise you and turn out to be an ideal big brother or substitute momcat. Although a companion cat is often just what a bored only cat needs, think carefully first. An only cat who has grown up alone, is more than three or four years old, has little or no experience with other cats, and is closely bonded to one or just a few people is likely to reject other cats. The experience will be traumatic for the new cat, plus you risk causing stress, illness, and long-lasting behavior problems in your cat. Think it through, know your cat—and follow your heart.

Quick Tips

1. The worst thing you can do about odd, quirky feline behavior is respond ambivalently or inconsistently. Tolerating a particular behavior pattern, then suddenly becoming upset about it (usually because your own needs or circumstances have changed) is a recipe for a very confused cat.

2. Don't seek instant cures for misbehavior. And don't bounce from cure to cure, trying one thing after another in increasing desperation. This invariably makes the misbehavior worse.

3. If your cat's hiding, ask yourself if he has a reason. Is there a new cat or new person in the house? Lots of stress and upset? A construction project? New carpeting? If so, leave him be. Make sure he's eating,

drinking, and using his litter box—and make sure he can get to every-thing he needs without venturing too far from his chosen hidey-hole. He'll come out when he's feeling more confident and secure.

4. Every cat has behavioral quirks. But if an unusual behavior shows up suddenly, or changes radically, schedule a visit to the veterinarian. Ill cats and cats in pain often show their distress with weird behavior rather than more obvious signs of pain or illness.

Chapter 10

Cat-versations

What Does *That* Mean?

It's easier to outsmart your cat if you know what she's talking about. Figure out what she means when she meows—or hisses or growls, trills or chitters, purrs or snarls. Is she begging for playtime? Complaining that the fresh catch of the day, isn't? Demanding that you move her favorite bed closer to the heater? Saying *I love you*? Telling you that there's a mouse under the stove? Or a fire in the basement?

Cat Sounds, Decoded

Among themselves, adult cats communicate through a subtle blend of scent, touch, and body language. In a magically efficient shorthand, the set of an ear, the swish of a tail, a gentle angling of whiskers, or the raising or lowering of eyes speaks volumes. A fleeting nose-to-nose "kiss"; a swift, polite sniff of a fellow feline's rear end; a cheerfully placid, side-by-side sidle; and a moth-soft paw-swipe all communicate more accurate and timely information than we humans could convey in hours of conversation.

What you'll seldom hear in a group of adult cats interacting with fellow cats, though, is *meow*. The meow's for *you*.

MEOWER POWER

Feral cats don't meow much. Wildcats get in most of their lifetime quotient of meows as kittens or cubs. Michael Owren of Cornell University's Psychology Department theorizes that domestic cats use meows as eminently practical ways of getting your attention, and then getting what they want. Over time, he notes, cats must have discovered that humans are very sensitive to pitch, and that the

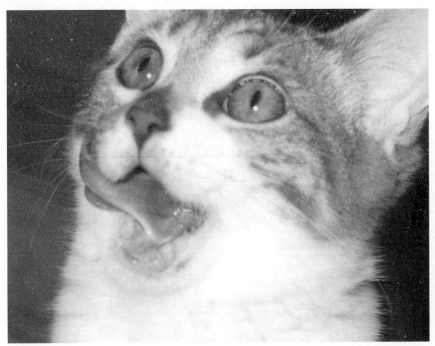

Treat, please, Mom! Jessie punctuates an irresistible request with a curl of her tongue.

pitch of a meow could have a strong emotional and motivational effect on them. Bingo!

Your cat's ancestors figured out long ago that these odd, upright cat-servants were primarily vocal communicators (and that humans are clueless when it comes to the nuances of scent and feline body language). So they learned to adjust their natural communication style to accommodate our preferences, and weaknesses. By closely observing the results of various meows, cats learned to use our sensitivity to pitch to get exactly what they wanted. The meow also reveals something of the history of our domestic cats, and of our relationship with them.

Over thousands of years, domestic cats have become, through evolution, self-selection, and selection by humans, neotenized. (See chapter 4.) When communicating with you, your neotenized cat speaks a language that in the wild is almost exclusively used between MomCat and her kittens: meows, chirps, trills, and kittenish squeaks. The meow language is primarily baby talk. But there's much more than the manipulative meow in your cat's vocabulary.

YOUR CAT'S VOCABULARY

In 1944 researcher Mildred Moelk published "Vocalization in the House-Cat: A Phonetic and Functional Study," the first attempt to enumerate, study, and document the sounds made by domestic cats, and to interpret their meanings. Moelk found that feline vocalization is a highly individualized art form. There are lots of similarities among cats, but each cat customizes her own dialect of catspeak.

For over five years, Moelk listened to, recorded, and classified the sounds made by her own cats, and recorded other cats' vocalizations "to check both sound and interpretation." She discovered that cats can make sixteen distinctly different sounds, of three types:

- Four *murmurs* produced with the mouth closed

- Six *vowel patterns* (*meow* and variants) produced with an open mouth that's gradually closed

- Six *strained intensity sounds* produced with the mouth held tensely open in a fixed position

HOW YOUR CAT MAKES SOUNDS

Unlike you, your cat uses both inhaled and exhaled breaths for vocalizing. By altering the force and speed of the air she inhales or exhales, and by opening her mouth wider or faster, your cat can instantly transform a quiet murmur of annoyance into a screeching wail of complaint or anger. Your cat's voice is produced by the vibrations induced in the vocal folds of her larynx by inhalations and exhalations. When you speak, you "shape" vowel sounds by changing the position of your jaw, the shape of your mouth, and both the shape and position of the tip of your tongue. Your cat does the same thing much more efficiently, simply by altering the tension of her throat.

Tabigail and her adopted daughter, Angel, trill and chirp to each other.

BABY TALK

Because your cat is a permanent kitten in many ways, it's only natural that she communicates with you using the mother–kitten vocabulary. When MomCat returns to her nest after hunting, or when she wants to summon her kits or encourage them to follow her, she calls out with a gentle trill or chirp. The kittens respond with kittenish squeaks and excited chirrups. Mom's greeting trill usually means safety, care, food, warmth, snuggling, and a nice bath. Your cat sees you as MomCat, so she greets you with trills, chirrups, mews, and squeaks when you get home, at playtime, and at dinnertime. This kitten language is remarkably rich and versatile.

MURMURS

Produced with the mouth closed, a murmur is composed of an initial voicing of an *mmm* sound during either inhalation or exhalation, followed by a trailing off into an *nnn* sound. The character and duration of the sounds in between determine the meaning of the murmur. A rush of breath can prolong and modulate the murmur, which can also include one or more little rolls or trills. The most familiar murmur, and everyone's favorite cat sound, is the *purr*, a continuous vibration produced during both inhalation and exhalation.

Purring can occur simultaneously with other vocalizations. It varies from loud to soft, rough to smooth, depending on the cat and the situation. Although it's often said that "a purring cat is a contented cat," the purr can indicate an overflow of *any* intense feeling: contentment, happiness, rage, or pain. Cats often purr while giving birth, during mating, when injured or dying, and at other times of high stress. Two-day-old kittens spontaneously begin to purr, to orient themselves and to signal to MomCat that they're happy and her milk is flowing. A subordinate cat may purr to show deference to a dominant cat, lessening the chances of a face-off, attack, or fight. The purr is a handy distraction technique, putting a potential attacker off guard or even evoking nostalgia for the peaceful days of kittenhood.

Mystery of the Purr, Solved at Last (Maybe . . .)

The actual mechanics of the purr were long a mystery. Thanks to sophisticated acoustic measuring techniques, it's now thought that the primary purr mechanism is "centrally driven periodic laryngeal modulation of respiratory flow." That

Purring Makes the World Go 'Round

When it comes to purring, international cats display astonishing diversity and creativity. Danish cats *spindle,* while their Dutch cousins *snorren* or *spinnen.* In Gaelic, cats *crònan* and *dùrdan,* while Hawaiian cats hum *non-olo.* In Japanese, *gorogoro* is a purr. In German, a purring cat says *schnur-ren.* Portuguese kitties *ronronar,* while next door, their Spanish cousins *ronronear* or *ronroneo.* In Finland, a purr is *kehrata.* In Italian, it's *fusa.* Welsh kitties hum *grwnan.* And in Hebrew, cats *geergoor.*

is, purring is controlled by your cat's central nervous system, and the actual vibrations are produced by modulation, or "gating," of respiratory flow by her larynx. Though it seems to radiate from her whole body, the purring sound emanates from your cat's nose and mouth.

Scientific understanding of the purr is still evolving. But one fact known since ancient Egyptian times is that purrs are good for *you.* Research shows that stroking a purring cat can lower your blood pressure and pulse rate and increase your subjective sense of peace and well-being. It's been suggested that the purr's vibration may help heal injuries faster. (See box on page 256.) Scientists from the Fauna Communications Research Institute in North Carolina found that domestic cats purr at a frequency of 27 to 44 hertz (cycles per second), about two or three octaves below middle C. Other studies have shown that exposure to frequencies of 20 to 50 hertz strengthens human bones and helps them grow.

So Nice to See You . . . Where's My Treat?

Other murmurs include a level or upwardly inflected greeting or request trill, often repeated in series. A closed-mouth request call, ranging from softly coy to demanding, is also often repeated continuously, with increasing pitch and volume, until your cat's request, or demand, is satisfied. Your cat follows this with a brief murmur of rapidly falling intonation that indicates confirmation or acknowledgment, signaling that she's gotten, or knows she's about to get, what she wants.

THE VERSATILE MEOW

With the music of her meows, your cat plays you like a violin. She sings her endlessly varied meows, in concert with kittenish squeaks, trills, and chirrups, to complain, beg, flatter, plead, and warn; and to express fear, boredom, irritation, bewilderment, disgruntlement, or excitement. You won't be surprised that many meows are complaints. Others are requests—from polite to forcefully demanding—for food, play, or attention.

The better your cat trains you to understand her individual meows, and the more you encourage and respond to them, the richer her vocabulary will become, the more frequent and meaningful your conversations will be—and the better your chances of outsmarting her are. As Moelk noted, "The house-cat's dependence on its hominid companions and educators encourages the use and extension of its vocalizing powers." Your cat might develop a specific, personalized meow to beg for play with a particular toy, or to demand that you shut your computer off *immediately*.

Jessie uses her versatile meow to ask for some playtime.

The Making of a Meow

In forming a meow, your cat usually opens her mouth gradually while her vocalization—beginning with the *mmm*-like sound—is already in progress. The strength and speed of her breath, and how rapidly she opens and closes her mouth, determine the volume, character, and intensity of her meow. Exactly how

In any language, cats are splendidly equipped to speak for themselves. In English, kitty says *meow* (or *miaou, maiow, meaow, meaw, meeow, mew, miaouw, miaow, miau, miauw, miaw, mieaou, miow,* or *mi-owe*).

In Arabic lands, cats say *naoua*. In China? *Mio* or *ming*. French cats *miauler* while their German cousins *miauen* and their Spanish friends *maullar*. In Greece, kitties say *larungizein*. Hungarian felines *nyávog*. Vietnamese cats *kêu meo meo*. Swedish kitties *jama*; Turkish cats *miyavlamak*.

Is Purring a Biomechanical
Self-healing Mechanism?

Because we so enjoy the soothing, flattering qualities of our cats' purrs, we tend to assume that they're purring just for us. But are they?

Why did evolution favor this seeming luxury, this particular sound-and-vibration medium, so pleasant and rewarding for humans and so convenient in intercat communication, in an animal that is, in its natural state, a solitary predator? Why should a hard-pressed cat—injured, under stress, or even dying—expend the precious extra energy necessary to produce a purr? Could purring provide some sort of survival advantage?

The research that uncovered the purr's mechanism offered a few intriguing hints. The "fundamental frequency" (number of vibrations per second) of the purr, averaged over inspiration and expiration of breath, is about 25 hertz (cycles per second), and ranges in a relatively narrow band between 23 and 31 Hz. Unlike the cat's other vocalizations (meows, for example), for which the fundamental frequency generally exceeds 100 Hz, purring is strictly a low-frequency phenomenon. The many purrs measured by the researchers varied greatly in loudness and duration, but very little in frequency. Purrs tend to be rich in harmonics, those higher-pitched "overtones" that accompany any sound and are generally multiples of the fundamental frequency. (For example, purr harmonics tend to be at 50 Hz, 75 Hz, 100 Hz, etc.)

An individual cat's purr varies in vibration as little as 20 percent, and seems to be highly stable and consistent over his lifetime, as well as independent of the cat's gender, age, size, and weight. Domestic cats, cougars, and cheetahs all purr within the same range of frequencies, varying mainly in the harmonics produced.

Purring is clearly meant for "up-close and personal" encounters. The purr's low fundamental frequency (25 Hz) cannot be heard farther away than about twenty inches. Indeed, this frequency falls below many humans' range of hearing altogether. When we savor the sounds of a purr, what we're hearing is mostly the harmonics. And even these can't be detected farther away than about ten feet.

Elizabeth von Muggenthaler, a bio-acoustician, devoted cat owner, and president of the Fauna Communications Research Institute in Hillsborough, North Carolina, has long wondered about the "why" as well as the "how" of the feline purr.

Using her knowledge of bio-acoustic measurement and analysis, she recorded and measured purrs in several species of cats, including caracals, servals, ocelots, lions, cheetahs, and domestic cats. Despite their differences in species, size, respiratory patterns, and other factors, all these cats purred at more or less the same fundamental frequencies, right around 25 Hz. In theory, this shouldn't happen. Every cat generated strong frequencies between 25 and

50 Hz. All the cats' purrs had a dominant frequency or a strong harmonic right at 50 Hz.

Von Muggenthaler had been looking at data compiled by both the Soviet space program and NASA, detailing research into certain ranges of vibrations that appeared to promote bone growth, increase bone density, and retard the loss of bone density experienced by astronauts living in zero-gravity environments. At the same time, she found numerous studies detailing how cats seem to heal much more quickly after injuries and surgery than other animals. Then a 1999 presentation by Dr. Clinton Rubin regarding harnessing vibrations to augment human bone healing gave birth to von Muggenthaler's exciting, and revolutionary, theory: The feline purr may be a biomechanical self-healing mechanism.

Rubin and his colleagues had found that exposure to frequencies between 20 and 50 Hz at low decibel levels "creates the robust striations of increased bone density." Upon further investigation, von Muggenthaler discovered hundreds of studies that supported the notion that vibrations of certain frequencies could not only promote bone growth and speed repair of fractures, but also provide pain relief, accelerate healing of tendon and muscle injuries, reduce swelling, improve joint mobility, and relieve difficult breathing. The feline purr fits beautifully into the heart of these "healing frequencies."

"Is it possible that evolution has provided the felines of the world with a natural healing mechanism for bones and other organs?" Von Muggenthaler asked. "Researchers at Fauna Communications believe so." In 2001 she presented "The Felid Purr: A Healing Mechanism?" at the 142nd annual Acoustical Society of America, American Institute of Physics, International Conference, and was also published and peer-reviewed in the proceedings from the 12th International Conference on Low Frequency Noise and Vibration and Its Control, held in Bristol, UK, in September 2006.

As highly efficient, solitary predators in the wild, felines enjoy the leisure of remaining sedentary, though alert, much of the time. Pursuing sudden hunting opportunities calls for instantaneous, sporadic bursts of vigorous chasing, leaping, and pouncing, a pattern that may well result in strain and injury to a usually sedentary animal's bones, muscles, and ligaments. A built-in mechanism providing pain relief and healing, and promoting bone growth—an "internal vibrational therapeutic system" that would kick in and work in the background while the cat rested between hunts—would provide a splendid survival advantage. The purring cat could maintain stronger bones and muscles, even while largely sedentary, and recover faster from any injuries he did suffer.

Research continues into the possibilities of transferring the healing effects of a purring cat to nearby creatures, including humans. Maybe a catnap isn't just a catnap after all: Maybe it's a physical therapy session. Next time you're under the weather, curl up with a purring cat. It can't hurt, and it might be just what the doctor ordered.

your cat begins her meow (with a closed mouth, partly open mouth, or inhaled murmur), plus the tone, duration, and inflection of the meow, helps determine its vowel pattern and meaning.

The demand meow varies tremendously, depending on what your cat wants, and what she thinks her chances are of getting it:

- Upward-inflected, coaxing meow that begins with a murmur

- Insistent, drawn-out, and single-minded begging meow

- "Cat whisper," a short, quiet request

Any of these can be combined and repeated, depending upon the urgency of the situation, and the degree of success (or failure) experienced. If your cat's expectations alter while she's actually meowing a demand (for example, you leave the room or ignore her), she may well segue into an uncertain, wavering, upwardly inflected meow of complaint or bewilderment.

The "Silent Meow"

It looks like your cat's meowing, but you can't hear a thing. Is the silent meow really silent? One thing's for sure: It's irresistible.

Your cat hears sounds of much higher frequencies than we humans or even dogs can detect—up to 100 kilohertz (100,000 cycles per second). We can hear only up to about 20 kHz in our prime and even less as we age. Along with the shape, mobility, and focusing facility of her external ears, this sensitivity allows your cat to detect extremely high-pitched or faint sounds, like the tiny squeaks of mice, the trills of kittens, and the silent meow.

To another cat, the silent meow is just an ordinary, high-pitched meow. But for most humans, especially cat owners, it's cute, cute, cute, immediately provoking the urge to pamper, pet, and please. Most cats quickly discover how effective this soundlessly plaintive display is in getting what they want, and they use it shamelessly.

Watch Out: And That Means You

Your cat produces *strained intensity sounds*—spitting, growling, the anger wail, snarling, hissing, and the mating cry—with her mouth held tightly open, and her breath forced through her tensed throat. Though they're usually directed at other cats, you might find yourself the target of these startling and often scary vocalizations. Growls, snarls, shrieks, and hisses can all indicate fear or pain. Never ignore these emphatic, meaningful sounds.

A startled cat—one who suddenly sees an enemy or potential threat, or who's grabbed from behind by an insensitive lout of a person—often spits, a quick, intense forcing of breath outward through her slightly opened mouth. A spit means she's ready for instant action: body tensed, claws at the ready. Startling a cat is a bad idea.

The growl originates deep in your cat's tensed throat. Her body language—crouched posture, lashing tail, mouth held stiffly and slightly open—underlines the intensity of this low-pitched, menacing sound. Because of the extreme tension in the throat, a growling cat may drool, or interrupt her vocalizing to swallow repeatedly.

Rival males growl face-to-face, each daring the other to make the first move. Females signal unwillingness to mate, and proclaim the end of a mating session, with an unmistakably annoyed growl. A chorus of growls can be petrifying, but growls are often just dramatic bluffs and taunts: *Back off, buddy! I'm one baaad cat!*; *Go ahead . . . make my day!*; *Hah! Who's gonna make me?*; *Yeah? You and what army?*

If a fight begins in earnest, a duet of anger wails follows the growling. An angrily growling cat gradually opens her mouth wider, producing a loud, extended wail, then gradually closes her mouth again to a low growl. As the fight escalates and intensifies, the anger wails speed up and become shriller and more piercing.

If a growling standoff doesn't lead to an actual fight, one or both of the parties may engage in a less fearsome wail that Moelk called angry protest, a sound of great, but not murderous, annoyance and warning. The volume, pitch, and duration of these intense sounds is meant to convey to all in earshot that the speaker is a formidably large, strong, and potentially dangerous creature, and not to be provoked or challenged.

During an extended, serious fight, the combatants may snarl—a loud, harsh vocalization characterized by a rapid, heavy inhalation with sudden termination. With luck, a sudden shriek or snarl can startle a disputant into loosening her grip, or even frighten her into fleeing the battlefield. The antagonists mix and mingle their snarls, growls, and wails in innumerable ways, making a chaotic, clashing cacophony. They may repeatedly back off and reengage. Because a fight between rival males involves such a large element of competitive display, the caterwauling of these midnight concerts can go on for hours, provoking much shoe throwing and neighborhood unhappiness.

HISSY FITS

Like a growl, a *hiss* is a threat, but a far less serious and intense one. Old folklore suggests that cats hiss to imitate the spitting of poisonous snakes. A hissing cat is wary, short on patience, and prepared to stand her ground. Depending on the situation, a hiss might mean she's annoyed, and maybe a bit scared, but intends to stand up for herself anyway. A hiss can indicate pain or stress, too.

A hiss definitely means, *Stop! Back off.* If your cat hisses while you're petting, grooming, or examining her, stop immediately. Try to determine the cause. Let her calm down. If she hisses whenever you touch a certain spot, she may be ill or injured. Schedule a visit to the veterinarian.

LOVE HURTS

The *mating cry,* said Moelk, is a "much modified form of the demand." Along with lots of swallowing, and licking of chops, the openmouthed female cat begins vocalizing while inhaling, pronounces an emphatic variant of a long, extended meow while exhaling, and lets her cry trail off as her mouth closes. The mating cry varies continuously, from very quiet to very loud and intense. The volume, duration, intensity, and persistence of the mating call is just one more reason to get your cat spayed.

DID YOU SEE THAT BIRD?

An odd variant of strained intensity sounds is the rapid, click-click *chittering* cats make while bird-watching (along with high-pitched squeaks and mews). Some observers feel it's an overflow of excitement, or frustration at being unable to get at the tantalizing prey. Or your cat might be mentally

Chitter, chatter, chitter: Ooh, I could get one of those so easily; chatter chitter, yum, yum, chitter . . .

rehearsing the specialized jaw motions of the killing bite, or performing an exaggerated, imaginative version of the bite itself. Oddly, cats watching rodents seldom chitter. It seems to be a bird thing.

Outsmarting the Too-quiet Cat

Is your cat the strong but silent type? Do you crave more cat-versation?

- When she meows, meow back—in her own language.

- Use a high-pitched, kittenish voice, especially at first.

- Observe and listen. Learn the nuances of her personal vocabulary.

- Talk to her, frequently: while serving meals, at playtime, while cuddling.

- Get into the habit of saying good morning, good evening, please, thank you, and excuse me to her at appropriate times. This will help teach her that conversation is a normal, desirable part of life with humans.

However much you practice your meows, you may never become fully fluent in cat. Still, you might get a chattier cat, it's great fun, and it'll help strengthen your bond. In vocalization as in so much else, your cat's an individual. Accept her vocalization preferences, and treasure those meows she chooses to share with you.

Outsmarting the Motormouth Cat

Perhaps you have the opposite problem: a cat who never shuts up. If you're annoyed at your constantly yammering cat, or frustrated because you can't figure out what she wants, it's time for two immediate steps.

First, schedule a visit to the veterinarian to rule out illness. Excessive vocalization, or any major or sudden change in vocalization habits, could mean your cat is ill.

Then take an honest look at your own attitudes and your relationship with your cat. Are you being fair? Have you given her plenty of time to adjust to changes in your life and home? Are you sure you haven't inadvertently introduced stress into her world? A new baby, new spouse, a move to a new home, even new furniture or carpeting can cause your cat immense stress and bewilderment, which can cause increased howling, crying, or incessant meowing.

Is your cat getting on in years? Perhaps her vision or hearing has deteriorated and she's having trouble figuring out where she is, and what's going on. Perhaps she's feeling lost, confused, and disoriented, common causes of howling and meowing, especially in elderly cats.

Cats often wail, cry, and howl after losing a friend, feline or human. Be patient with your grieving cat, and be sure she continues to eat and drink normally. A bereaved cat can become deeply depressed, which can quickly lead to serious illness.

And . . . aren't you being unreasonable to expect a naturally curious, friendly animal to remain silent as a robo-cat? Tune in and listen. She might be trying to tell you something very important.

CUTTING THE CHATTER

Some breeds, especially Siamese and other Orientals, can drive their otherwise loving owners mad with their incessant, piercing meows, howls, and wails. Here's how you can help cut the chatter:

- Rule out illness as a cause.

- Start ignoring her vocalizations: When she meows, don't talk to her, play with her, feed her, pet her, let her in or out, or anything else. This can be very difficult, especially at first. But *any* feedback is likely to reward and encourage her vocalizing. You *must* be consistent. By responding (even yelling at her to knock it off), you inadvertently reward her unwanted behavior.

- Observe the situations and circumstances in which she's most likely to meow. Distract her *before* she starts meowing with a play session or cuddling.

- Watch for opportunities to reward her when she's especially quiet with treats, extra playtime, or whatever she likes best.

- If your cat still can't seem to shut up, ask your veterinarian about prescribing a short course of an anti-anxiety medication. This can help get her (and you) over the hump while you work on modifying her behavior.

- Ask yourself if the problem might be *you*—not your cat. (Maybe it's *you* who needs the anti-anxiety medication.) Adjust your attitude.

Cats around the World

When addressing an international cat, it's only polite to do so in her own language. Here's how to say "cat" in many lands:

Arabic: *kitte, otta, qit*

Armenian: *gatz*

Basque: *catua*

Chinese: *miu, mio*

Danish, Dutch: *kat*

French: *chat*

German: *katze, katti, ket*

Greek: *catta, kata, ga'ta*

Hawaiian: *owan*

Hebrew: *chatul, hatul*

Icelandic: *köttur*

Indonesian: *kutjink*

Italian: *gatto*

Japanese: *neko*

Latin: *felis, catus*

Mohawk: *tako's*

Norwegian and Swedish: *katt*

Polish: *kot, gatto*

Romanian: *pisicâ*

Spanish, Portuguese: *gato*

Swahili: *paka*

Turkish: *kedi*

Welsh: *kath*

Yiddish: *ketzel, ketzela (kitten)*

Why work so hard to try to shut off what could be a vital and interesting line of communication, though? Listen to your cat. Discover what interests her. Talk with her about it. You may find a new world opening up to you, and feel closer than ever to this mysterious, marvelous creature.

Quick Tips

1. Cats are highly attuned to high-pitched sounds. They pay more attention to the squeaks of kittens (and mice) than lower-pitched sounds. If you want to get your cat's attention, speak (or meow) in a high-pitched voice.

2. Is your cat getting on in years? Like us, cats start to lose their hearing in the upper ranges first. If it seems your cat is having trouble hearing you, try speaking to her in a lower-pitched tone.

3. Learn the difference between a growl and a purr—it can be quite subtle. But the meanings are as different as lightning and a lightning bug.

Chapter 11

Shedding: Cat Hair, Everywhere!

It's a myth that dark-haired cats shed only on light-colored surfaces while light-colored cats prefer navy blue suits. All domestic cats shed, everywhere and nearly continually. (An exception is the Sphynx, a hairless pedigreed breed. They leave grease spots.) Shedding is a natural function of a healthy cat.

Your cat's coat goes through cycles: growth, transition, and a resting phase. Shedding happens during the growth phase, when new hairs push old ones out of the skin—onto your sofa or suit.

For wildcats, major shedding, called molting, is a twice-yearly event, brought on by changing light levels in spring and fall. It's the number of hours of sunlight per day (the *photoperiod*), not temperature, that determines how much and when cats shed. The more exposure to light, the more frequent the shedding,

Hazel-Marie, like most cats, is a fastidious self-groomer.

and the greater the amount of hair shed. Most domestic cats, especially indoor cats, shed throughout the year as their coats cycle through the three phases.

Some cats shed very little, while others seem to exude a constant cloud of fluff and leave furry mats and tufts

Calico Petunia has a hair color to contrast with every household surface.

everywhere they rest. Cats with short, coarse-furred coats usually shed less than those with long or fine hair. But even short-haired cats can have dense, fluffy undercoats that leave memorial traces everywhere. Coarser fur, once shed, tends to settle down in one spot and is relatively easy to scoop up. But fine, silky fluff wafts about in clouds before attaching itself to a variety of fabrics and other surfaces with astonishing tenacity, often seeming to weave itself right into the nap.

(There's also the curious matter of pure black cats who shed white fur on dark fabrics, and all-white cats who contrive to adorn your dress whites with black fur—but these are cosmic mysteries best left to philosophers and metaphysicians.)

Outsmarting Cat Hair

One of the easiest ways to minimize excessive shedding, and guarantee your cat's good health, is to feed him a high-quality diet. A balance of essential fatty acids, vitamins, and minerals is essential for a lush, healthy hair coat.

As you peruse cat catalogs and websites, you'll frequently see various supplements touted as necessary to achieving nutritional balance. But if you feed your cat a high-quality, complete, balanced commercial food made by a reputable manufacturer and approved in feeding trials by the American Association of Feed Control Officers (AAFCO), your cat will get everything he needs for a healthy coat. If you think your cat needs supplements to control shedding or improve his hair coat, ask your veterinarian.

The easy answer to "cat hair, everywhere" is "grooming, everywhere." Keep a flea comb (the kind with narrowly spaced teeth) and a cat brush (or child-size hairbrush) handy next to wherever you sit down with your cat to relax—by your

Long fur or short, all cats shed. Chrysanthemum's favorite climbing tree regularly yields enough fluff to build a couple of new cats.

easy chair or where you watch TV, for example. Keep extras handy around the house. Take advantage of odd moments when your cat is on your lap, or just standing around, to get in a few minutes of combing and brushing. Every little bit helps. This casual, fun approach is especially good for fussy, impatient cats, and those who dislike long formal grooming sessions.

Even if your cat is grooming-resistant, try to get in at least a few licks every day. Don't wait until he has mats. These hard knots and clumps of hopelessly tangled fur tend to form in a cat's underarms (on all four legs). Cats with long, fluffy fur or thick, soft undercoats are particularly susceptible to mats, but any cat can get them. Mats can be very painful, pulling and tugging against a cat's tender skin and making it difficult for him to groom or even move properly. They can also trap moisture and bacteria next to the skin, leading to infections and skin irritation.

If your cat's badly matted, don't try to break up the mats yourself. Never try to cut out mats with a pair of scissors; you could quickly cause severe injury. Feline skin is

Cats spend about one-third of their time self-grooming. A cat who seriously neglects his grooming, or stops grooming altogether, may be ill or in pain. Schedule a visit to the veterinarian without delay.

very delicate and stretchy, and cats being groomed are notoriously wiggly. Instead, make an appointment with a reliable professional groomer. Ask your veterinarian for a recommendation. One of the clinic's technicians may offer grooming services.

Hairy Cleanups

Here are some of the easiest tools to remove shed feline fur from your clothing and furniture, and directly from your cat.

- Your dampened hand. Run it over the hairy surface (cat or upholstery). You'll capture copious quantities of hair with little effort.

- A dry or slightly damp common household sponge.

- A hard-rubber, triple-bladed squeegee-like device designed to remove cat hair (available from pet supply retailers and catalogs).

- An electrostatically charged duster. Use this on non-upholstered furniture, hard floors, electronic equipment, and other household surfaces that accumulate cat fur.

- A velvet brush, sold for removing lint from clothes.

- Lint balls. These are Ping-Pong-ball-size balls covered with what looks like the "hook" surface of Velcro. Toss several of these in the washing machine with your laundry. They capture lint, cat hair, and other fuzzies.

- Rollers coated with a layer of sticky paper that picks up the hair. The used portion of sticky paper is then stripped off, revealing a fresh surface ready for action. If you have a lot of cat hair to pick up, you'll go through a lot of refills.

- Ordinary packing tape works almost as well as a sticky roller.

- Rollers made of a rubbery material that attracts and picks up cat hair (like a sticky-tape roller), but never needs refills. Rinse off the hair and the roller's ready for more.

- A grooming glove, covered with small rubber nodules, that strips loose hair directly from your cat as you stroke him. Many cats adore these. It feels like a glorious massage-and-petting session, not grooming—great for fussy, impatient cats. It works best on short hair, and is useless for serious tangles and mats.

NOTES ON VACUUM CLEANERS

Vacuum cleaner attachments tend to get clogged with fur quickly, especially on carpet and upholstery. Your vigorous vacuuming merely forces fur deeper into the fabric. Stop frequently to clear fur from the attachments.

To loosen hair: Before vacuuming, mix two tablespoons of liquid fabric softener in one cup of water in a spray bottle. Spray the upholstery, let it dry for a few minutes, then vacuum. This works well on carpeting, too.

What about using a vacuum cleaner to remove loose hair directly from your cat? Sound ridiculous? Pet supply retailers feature vacuum cleaner grooming tools made just for use on pets. There are apparently cats who not only tolerate this, but actually seem to enjoy it. If you feel your cat might be one of these elite few, get a small, fairly quiet vacuum cleaner and give it a try. But wise owners won't insist.

Cat Hair and Computers

If you use a tower-style desktop computer, and if your cat has access to your computer room, your computer's innards are undoubtedly coated with a layer of fur. The more cats, the more fur. While this isn't usually a huge problem, over time the buildup can cause electronic components to run hotter than they should. Enough hair can clog cooling fans and air intakes, and interfere with connections, causing sporadic (and maddening) malfunctions. If you use a laptop or other small portable device, cat hair can get inside, too, but cleaning out these types of devices is definitely a job for a professional. Every once in a while, take your laptop to a computer repair shop for a complete cleaning.

For your home tower computer, though, you can do it yourself. Every once in a while, do a:

C(4): COMPLETE COMPUTER CAT-HAIR CLEANUP

- Escort all cats out of the room, and close the door.

- Shut down all computer equipment.

- Turn off all power and unplug all computer components.

- Disconnect all components. Even if you think you'll remember how to reconnect everything (I never do, and I was a software engineer for years), draw a diagram of how everything is connected before you take it apart.

- Place your main computer case (tower or other case) on a convenient, well-lighted work surface, such as a bench or table.

- Consulting your system's documentation if necessary, carefully open your computer's case. (This is usually harder than it should be.)

- There are two theories for cleaning computer innards: *blowing* and *sucking*. Either can be accomplished with a small, plug-in canister-type vacuum cleaner and a set of special "miniature" attachments, designed for cleaning electronics and delicate items, *if* the vacuum has a "blowing" port. (*Note: No battery-powered vacuum cleaner is powerful enough to extract accumulated cat hair from electronics.*)

- *Blowing* out the dust and hair is considerably messier, but some computer experts find it safer for the electronics.

- *Sucking* is much neater, but it requires more care and a lighter touch. If you choose this approach, use the tiny attachments to suck out accumulated dust and hair. Don't bump, dislodge, or disturb connections or components. "Float" the attachment over the circuit boards and other electronics, rather than actually touching them.

- A combination approach, carefully sucking most of the visible hair, followed by a final blowout, works well and minimizes airborne dust.

- Pay special attention to the fans and intakes, which trap a lot of hair.

- Put everything back together carefully and methodically. Consult your diagram. Don't rush.

- Vacuum your keyboard and mouse. Use your miniature crevice tool and brush to get down between the keys.

- Power all components back up and test. Make sure your system still boots up properly and operates normally.

Warning: If you feel at all uncomfortable taking your computer case apart like this, or if your setup makes it physically difficult, call a computer service professional. These technicians can come to your home or office to clean and service your equipment, or you can drop off your computer at their shop for cleaning.

- Welcome your cats back into your computer room to begin the hair-accumulation project anew.

If your cats share your work space like mine do, you'll probably find you have to vacuum your keyboard quite frequently—at least once a week.

Cat hair can also clog other electronic gear: printers, scanners, phones, fax machines, and so forth. Use your miniature vacuum attachments to clean your equipment frequently. If you still use an old-fashioned CRT-style computer monitor, and your cats like to perch atop it (I call my big tuxedo Wizard, my "Monitor Wizard"), cat hair has doubtlessly been wafting through the venting grilles into the innards there, too. The outside is another job for the vacuum cleaner. But *never* open up or try to take apart your computer's monitor. That's a job for professionals only!

To outsmart a persistent monitor sitter, treat yourself to a fancy new flat-panel display. Only the most graceful (and stubborn) cats will try to colonize the top of one of these.

For dealing with other cat–computer issues, see Outsmarting Keyboard Tappers and Outsmarting Monitor Perchers in chapter 12.

Problem Shedding

Some amount of shedding is normal. But if:

- your cat is shedding unusual amounts of fur

- the amount he sheds has suddenly greatly increased

- his fur is coming out in clumps and bunches

- he's scratching and tugging out large amounts of hair

schedule a veterinarian visit right away. Abnormal hair loss, even without itching, can signal illness or nutritional deficiency. Take along the packaging from your cat's usual food. Your veterinarian will want to check its nutritional composition.

IS IT ALLERGIES?

If the hair loss seems to be caused by your cat's scratching and worrying of his fur, he may be suffering an allergic reaction. Cats can be allergic to numerous household substances, including many common cat food ingredients. See chapter 6 for tips on outsmarting allergies in cats.

IS CAT HAIR MAKING YOU SNIFFLE AND SNEEZE?

For a cat lover, allergies to cats can be a nightmare. The sniffling, sneezing, rashes, itching, hives, teary eyes, asthma attacks, and general misery can degrade your family's quality of life, lower disease resistance, interfere with normal sleep patterns, and reduce productivity at work and school. In some cases, a severe allergy may even be life threatening.

In the case of a serious or life-threatening allergy, especially in a child, elderly person, or immunocompromised person, your only recourse, unfortunately, may be to find another home for your cat. Happily, though, this extreme measure is seldom necessary. With some planning, work, and investment in appropriate technology, even sensitive members of the family can usually coexist happily, safely, and healthily with the family cat.

It was long assumed that cat fur or hair precipitated allergic reactions in sensitive people. Then, however, it became apparent that a Sphynx (virtually hairless) cat could cause as severe a reaction as a superfluffy Persian. The next candidate for allergy culprit was inhaled cat dander, tiny particles of feline skin or hair that wafted into the air. But the real menace is a protein called Fel d 1 (*Felis domesticus* allergic 1), present in your cat's skin and saliva.

As your cat licks and grooms himself, he deposits and redistributes this protein all over his coat, and sheds microscopic particles of it into the surrounding air and onto surfaces in his environment. Every cat, regardless of breed or coat length, seems to be unique in the amount of Fel d 1 he produces and sheds, so there's no obvious way to choose a nonallergenic cat.

You're sniffling, itchy, and miserable, but you love your cat and wouldn't think of parting with him. What can you do to outsmart your allergies?

1. Make your bedroom a "cat-free zone." If you love sleeping cuddled with your cat, this may take some adjustment. Many sufferers find that just banning the cat from their bedrooms turns a big problem into a small annoyance.

2. If you work at home, consider the same strategy for your home office.

3. Invest in a true-HEPA (high-efficiency particulate air) cleaner for your bedroom. These systems filter out an estimated 99.97 percent of all airborne particles of 0.3 micron or larger—including Fel d 1 and a whole host of other irritants, germs, and pollutants.

4. Invest in a vacuum cleaner with a HEPA filter and filtered bags to maximize the effectiveness of your housecleaning efforts. Vacuum often. This will isolate the Fel d 1 in the bag and prevent it from spreading. These appliances can be expensive, but the potential health benefits for your family are enormous.

5. Minimize the presence of carpeting, drapes, and upholstered furniture as much as possible in your home. Fel d 1 is a relatively sticky protein, and tends to collect wherever dust collects. Instead of carpeting, use washable throw rugs, and launder them frequently in hot water. See chapter 5 for more reasons to replace wall-to-wall carpeting.

6. At each of your cat's favorite lounging spots, provide a washable soft towel, old blanket, small throw rug, cat bed, or even a couple of threadbare sweatshirts. Every few days, remove these and launder them in hot water. Your cat will appreciate the clean, soft personal beds, and the Fel d 1 particles will stay relatively confined.

7. Feeling brave? It's been reported that bathing a cat weekly in plain water removes most, if not all, of the accumulated Fel d 1. This isn't a popular program with most cat owners (not to mention their cats).

8. Not so brave? Most cats will stand still for a daily rubdown with a clean, damp washcloth. There are also a number of commercial products and premoistened wipes available that promise to remove Fel d 1 from your cat.

9. If you suffer from cat allergies, wash your hands thoroughly after touching him. Avoid touching your mouth or face while handling your cat.

10. If you suffer from cat allergies, delegate cat-grooming chores to another family member, or consider hiring a professional groomer.

11. Ask your physician or an allergy specialist for information about available over-the-counter and prescription (nondrowsy) anti-allergy medications. These can be quite effective for mild to moderate allergies.

12. Are you a really tough case? Ask your allergy specialist about desensitization treatments. Although the traditional treatments can be expensive and time consuming (they usually require repeated office visits over the course of several months), some former sufferers report excellent, permanent results. New vaccines on the horizon promise reduction of the desensitization period to weeks instead of months.

Outsmarting Hairballs

Cats invented extreme sports long before humans. Consider the ejection of a hairball. Unlike ordinary, everyday shedding, in which just a paltry few hairs are deposited here and there, a hairball presents your cat with an opportunity for a dramatic piece of feline theater: shedding a relatively large amount of hair in a daringly conspicuous fashion, and all at once.

The hairball arrives with great ceremony (the elaborate, staggering, backward walk), accompanied by dramatic sound effects (that unearthly moan-wail that precedes ejection), and in a carefully selected venue (Oriental carpets are preferred, the more expensive the better; shag carpets are also fine; but anywhere the result might be stepped on while still wet is acceptable).

Besides a feline art form, what *is* a hairball? A *trichobezoar* (the technical term for a hairball) isn't a ball at all, but a compressed, sausage-shaped mass of hair and saliva that can form anywhere from the back of your cat's throat to his anus. It usually passes uneventfully through your cat's digestive system. Depending on where the mass forms, it can exit either fore or aft.

Just about any cat, long- or short-haired, can develop hairballs. As he grooms himself, the rearward-facing barbs on his tongue strip off dead and loose hair and fur from his coat, and force him to swallow it.

Hairballs are normal and usually no big deal. All cats—from lions and tigers to bobcats and domestic cats—get them. That doesn't mean they're desirable. Large, frequent hairballs are no fun for your cat. Hair wads can obstruct his esophagus. Excess hair passing through his intestines can irritate tender tissues, resulting in small quantities of blood or mucus in his feces. Help your cat avoid or minimize hairballs by brushing and grooming him regularly to remove loose hair.

DEADLY HAIRBALLS?

Every time you step on a hairball, say a little prayer of thanks to the Cat Goddess. A big fat hairball on your carpet is a lot better than the same hairball inside your cat. Occasionally, hairballs can kill.

Signs of a serious hairball crisis:

- Inability to defecate; straining and crying in the litter box

- Loss of appetite; refusal to eat or drink

- Lethargy, depression, crouched posture

- Continuous or repeated retching or gagging

- Dry, heaving cough

- Swollen, tender abdomen

Any of these signs can mean an emergency; in combination they're definitely bad news. Call your veterinarian or the emergency veterinary clinic at once, and be ready to describe your cat's symptoms in detail. If you're lucky, administering a few doses of a commonly available hairball preventive can get things moving again. These nondigestible, goopy, fatty preparations are flavored to appeal to (most) cats. The greasy goop helps lubricate the cat's digestive tract, enabling those masses, clumps, and wads of hair to pass through more easily.

Sometimes a cat's digestive tract can become badly obstructed with little or no warning. Be observant, especially if your cat has had hairball problems before.

Even a short-haired feline can accumulate enough hair in her stomach and intestines to cause serious problems. All that ingested fur can form a fuzzy lining on the inside of her stomach, trapping acidic gastric juices and leading to painful ulcers. Over time the

Acclaimed artist Sir Sterling Silver notes a critic's reaction to his latest work.

stomach lining can be completely eaten away. A particularly dense or thick wad of hair can get trapped in the intestines, where it combines with feces and becomes impacted, causing painful constipation. Many cases of feline constipation are caused by impacted hairballs. Or the mass can form a complete blockage: a life-threatening crisis. Once the intestines are blocked, gas buildup can cause a fatal rupture. Surgical removal of the blockage is often necessary.

Sound scary? You bet! But fortunately, lethal complications from hairballs are rare. Most hairball problems are minor and transient, easily cured with a few doses of hairball goop.

A cat with chronic hairball problems may have a rough, scraggly coat and a depressed appetite, a low energy level, and poor overall condition. After ruling out other medical problems with a complete checkup, help her out by:

- Frequently brushing and combing.

- Feeding a high-fiber diet or nutritional supplements. Try adding a teaspoon or two of canned pumpkin (no spices), plain bran, powdered psyllium, or other fiber supplement to your cat's regular chow. Ask your veterinarian for advice and recommendations.

- Trying one of the new commercial diets formulated to help prevent hair-balls. Ask your veterinarian for recommendations.

- Regularly administering a hairball preventive. Overuse of these products can cause trouble, too, though. Ask your veterinarian about the best prod-uct, schedule, and dosage for your cat.

- Staying watchful for recurrences.

WHAT IF IT'S NOT JUST A HAIRBALL?

Too often, cat owners blame common symptoms—vomiting, poor appetite, leth-argy—on hairballs when the cat is suffering from a more serious medical problem that needs attention. Vomiting, retching, chronic diarrhea, reduction in eating or drinking, lethargy, or depression—especially in combination—call for a complete checkup as soon as possible. Excessive vomiting, in particular, can cause a seri-ous electrolyte imbalance that can have serious health repercussions for your cat.

Never assume that all that vomiting is "just another hairball."

Quick Tips

1. Outsmart your shedding cat by taking advantage of an odd but familiar feline quirk. Every cat owner has seen it: Cat enters a room that's virtually empty, except for an empty box, a dropped towel or article of clothing, even a scrap of cardboard. Feline makes beeline for that island as if it were the most attractive spot on earth. Use this feline peculiarity to guide your cat to spots you'd prefer he nap (and shed). Toss a fluffy towel, throw, or small rug on a chair or sofa, or in a sunny spot on the floor. When it gets hairy, replace it with another and launder it (a great way to recycle old towels).

2. Keep a supply of old towels, or really ugly towels bought on sale, for your cat's use. Launder and store them separately from your own towels.

3. Provide an irresistible cat bed, or several beds in locations your cat likes to lounge or nap. Beds made of polyester fleece (such as Polarfleece) or Berber fleece are particularly prized by comfort-seeking cats. Offer open, cup-shaped beds and cozy, private, cave-like beds. Place them both on the floor (near a heating duct in winter), and on higher surfaces. Notice which locations your cat prefers. If her own lounging facilities are sufficiently appealing, she's less likely to shed on yours.

4. Use machine-washable slipcovers. Keep some different styles and colors handy, and you can always make your furniture look brand-new at a moment's notice—or just gratify a whim to redecorate on the cheap.

5. Provide specially designed cat furniture: climbing trees equipped with carpeted ledges, fleece hammocks, and hideaways. These lofty perches are often more attractive to cats than human furniture that's closer to the floor. Cats feel more secure snoozing high up, where they can see who's coming and what's going on without worrying about being bothered or surprised. See chapter 6 for more tips on climbing trees and cat furniture.

6. If you can stand to ignore them for a few days, hairballs are much easier to clean up when they've dried out. Let the hairball dry, then just pick it up and toss it in the trash. Vacuum up any residue. If the hairball left a stain, use a cat-safe enzymatic stain remover on the spot.

7. If the hairball is particularly messy, is combined with cat vomit, or you're expecting the governor for dinner, clean it up immediately. The mess may contain hydrochloric acid and quickly and deeply stain carpets. (The same goes for plain cat vomit.)

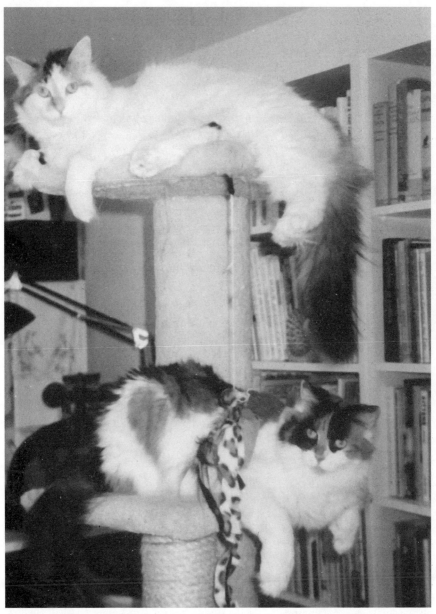

Calico sisters Angel and Chrysanthemum delight in one of many custom-enhanced cat trees available for their comfort and entertainment at Bobcat Mountain Studios.

Chapter 12

COOPs (Cats Out Of Place)

"No animal should ever jump up on the dining-room furniture unless absolutely certain that he can hold his own in the conversation."

—Fran Lebowitz

I take a laissez-faire attitude toward my cats' lifestyles and activities, and I have a rather more tolerant attitude about what I consider a Cat Out Of Place (COOP) than some owners might. Since it's my decision to keep my cats indoors (although there are numerous benefits for them, too), I feel it's my responsibility to help make their lives as interesting, stimulating, satisfying, and trouble-free—as wildcat-like—as I can.

If you don't approve of this sort of thing, set kitty ground rules, make sure the whole family agrees, and above all, be consistent!

Outsmarting Counter Cruisers

Many cat owners object to counter cruising (and dining room table incursions) on hygienic grounds. Cats are extremely clean animals, and all the cleaner if their litter facilities are well maintained. Cat paws are probably more sanitary than the sponge or rag you use to wipe down the counter. If you do allow your cats counter privileges, vigilantly guard the stovetop whenever it's in use. Don't leave the room, or turn your back—even for a second.

I don't mind my cats cruising the dining room table and kitchen counters. And since I very seldom use the stovetop (microwaving is more my speed), safety is less of an issue. It's a personal decision. Whatever you decide, set your ground rules and *be consistent.* If one person in your family allows your cat on the counters and another doesn't, that's going to be one confused, stressed cat!

Keeping your cat off the counters when no one's around can be a challenge. Here are some ways to outsmart a counter cruiser:

1. Offer a more interesting and attractive alternative, like a tall climbing tree, nearby, perhaps near a sunny kitchen window. It offers a better view and more luxury than the countertop.

2. Make the countertop an unattractive walking surface:

   • Place sheets of aluminum foil on it.

   • Place a piece of plastic carpet runner (spiky side up) on it.

   • Lay contact paper (sticky side up) or a double-sided tape product like Sticky Paws on inexpensive plastic place mats and lay these on the countertop.

3. Don't defrost a chicken or roast on the counter while you're away.

4. Don't leave anything edible or otherwise cat-attractive on your counters—not even a glass of water that can be tipped over.

5. Never leave knives, scissors, breakable objects, or other accidents-waiting-to-happen on your countertops.

6. If there's a window with a cat-attractive view, close it off with a shade or curtain.

7. Use booby traps to discourage incursions. (Before you do this, though, reread "'Nice' Matters" in chapter 4.) Fill several soda cans with coins

or pebbles and tape them shut. Pile these in an unstable pyramid near the edge of the counter where they'll fall down noisily if your cat jumps up.

Outsmarting Trash-can Tippers

Trash-can-tipping cats make a mess. They might also snack on garbage or other disposable items, risking injury, poisoning, and illness. Keep all trash and garbage away from your cat, and your cat out of trash cans and garbage bins. Here's how to outsmart your feline garbage-hound:

- Use trash and garbage cans with lids. Look for trash containers with swinging, snap-on lids, spring-loaded lids, or containers equipped with foot pedals for opening and closing.

- Keep your trash and garbage cans inside kitchen and bathroom cabinets. Obtain one of those clever mechanisms (available at home improvement centers) that attach to a cabinet door and roll the trash can out when you open the cabinet, then back in when you close it.

- Keep trash and garbage pails in a cabinet fitted with child-proof latches.

- Don't throw away food or spoilable trash in any trash can except the one in the kitchen.

- To discourage tippers, place a small cinder block, or three or four common bricks, in the bottom of each trash receptacle before inserting a paper or plastic bag liner.

- Trash or garbage that would be dangerous for your cat if she gets into it should be put outside (into a lidded outdoor garbage can) immediately.

Outsmarting Plant Chompers

Some cats just can't resist plants: live or dead, real or artificial, fresh or dried, toxic or benign. Protect your plant-loving cat from harm, and protect your (cat-safe) plants from her toothy depredations.

Dandelion adores tasty spider plants. But do they adore her?

- To make your collection of plants easier to protect, display them in groups. They'll look more lush and be easier to tend. It's also easier to teach your cat to avoid one or two specific areas than to keep her paws off all those tempting individual plants.

- Keep your plant collection in a cat-free zone, and call it the conservatory.

- Dried arrangements can include plants treated with toxic chemical dyes or preservatives and can be as dangerous as toxic live greenery. Sharp dried stems can injure your cat's mouth and internal organs. Display dried arrangements well out of your cat's reach—or not at all.

- Silk flower arrangements are decorative, last longer than live plants, and require no care but occasional dusting or rinsing. Your cat is much less likely to show interest in them, and they're certainly safer than live or dried plants.

- Sprinkle the soil in your plant pots, or the area where you group your plants, with something your cat finds repulsively stinky: pepper, Tabasco sauce, citrus rinds, or cotton balls soaked in flowery perfume.

- Keep your plants up off the floor. Use hanging planters, or place planters on (stable, nontippable) pedestals or narrow-topped tables with no landing zone for curious, leaping cats.

- Is your plant chomper just bored? When she heads for the greenery, distract her with a play session, or divert her to a game or cuddling session in a different room.

- Set up a bird feeder outside a window. Most cats find that watching the birdies is more addictive than other indoor pursuits, including greenery chewing.

- Provide her with her own greenery: pots of kitty grass. (See chapter 8.)

- Sacrifice a cat-safe plant or two. A couple of spider plants offer plenty of chomping fun. Your sacrificial greenery may look ratty after a while, but spider plants are nearly impossible to kill. Switch them periodically into a cat-free zone to recover.

Keep unsafe greens completely out of your cat's reach. Many common houseplants are irritating, toxic, or deadly to cats. The safest course is to get rid of them. But if you're especially fond of risky plants, keep them completely inaccessible to a curious, leaping, climbing cat. Grow them in a cat-free zone, or hang them securely from ceiling hooks. Pick up fallen foliage right away, and never let your cat drink spilled or overflowed plant water.

Outsmarting Furniture Loungers

It's easier, safer, and less stressful (for people and cats alike) to avoid putting temptation or attractive nuisances in your cat's way. I think it's a mistake to declare certain objects that are physically accessible to your cat off limits. Any chair I can sit on, so can my cats. They live here, too.

Instead of aversive stimuli, or constant nagging, use cat-free zones. Store or display items that are precious, valuable, fragile, or just too tempting to your cat in rooms or areas physically inaccessible to her. If you're mildly allergic to your cat, or would rather not have her crawling all over your lap, desk, or computer keyboard while you're working, make your home office a cat-free zone, and display your cat-tempting items there. Knowing that your Ming vase, or collection of first editions, or Grandma's brocade rocker is safe will reduce your stress, and your cat's stress, immensely.

If you feel you must keep your cat away from certain pieces of furniture or other items, and you can't lock these away from her, try safe aversive techniques:

- Balloons

- Booby traps (stacked soda cans filled with coins or pebbles)

- Proximity-activated voice recorders ("Get off the sofa!")

- Proximity-activated ultrasonic alarms

- Electric scat mats

Before considering these, though, reread "'Nice' Matters" in chapter 4.

If your nightmare is cat hair, everywhere, use slipcovers, washable throws, cat beds, and towels. See chapter 11 for more tips on outsmarting a shedding cat. If your problem is claw damage to your furniture, the tips in chapter 6 will help you outsmart your scratching cat and divert her to acceptable alternatives.

Outsmarting Bookshelf Scalers

As a benefit of walking upright, we humans view our surroundings from a relatively lofty height. We sometimes forget that our cats don't share this advantage. Because of their evolutionary heritage as tree dwellers, though, cats think and live in three dimensions. They not only *adore* heights, they seem to *crave* heights. It's only from up above that these small animals can spy on whatever's going on without being noticed, or just catnap above it all. High up,

The Wizard of Bobcat Mountain is a voracious reader. He prefers detective yarns with suave, tough feline heroes. (Don't tell Silver and Tabigail, but he's also a secret fan of romance novels.)

Soft Landing?

Your cat has an almost magical ability to flip over in midair and manage to land safely on her feet—if she has enough time.

Let's say your cat finds herself falling—from a tree, or the top of a bookshelf. Instantly, her brain sends a series of messages describing her current orientation in space and the positions of her limbs, body, and head to the balance mechanisms (vestibular organs) in her inner ears, and to her eyes. Now that she knows which way she's facing, she can turn her head and front limbs toward the ground. Using her tail as a balance beam, she aligns her spine, twisting it as necessary, with the rest of her body. At the same time, she spreads out her four legs, forming a parachute. Her flexible shoulder and leg joints, working together with her floating collarbones, give her a bit of extra bounce, while paw pads help cushion her touchdown. This extraordinarily complicated series of maneuvers all happens in less than a second, and in as little as two vertical feet of fall.

Still, falls are nothing to sneeze at. Though cats have survived some spectacular falls, falls kill or badly injure many cats. An awkward landing or a fall from a great height can result in severe internal injuries, a broken jaw, skull fracture, or smashed teeth. Veterinarians have a name for what happens to cats who fall from unscreened windows and city balconies: feline high-rise syndrome.

Cats can land on their feet—if they have enough time.

A short but awkward fall can cause serious injuries, too, because the cat's righting reflex might not have had time to kick in. Whether the cat walks away from a fall, or is badly injured, depends on several variables, including her age and physical condition, the height she falls from, how awkward or graceful the fall, and where and how she lands.

Your cat's righting reflex is a built-in emergency lifesaving procedure, not a party game. Don't "test" it or demonstrate it. At best, you'll insult, embarrass, or annoy your cat; at worst, you'll injure or kill her.

they feel secure, knowing that they'll get their best and earliest look at threats and opportunities.

Instead of yelling at your bookshelf-scaling cat (which won't do any good anyway), give her safe, acceptable access to lofty realms all her own. In chapter 6, you learned all about climbing trees. To outsmart your bookshelf-scaling furry Tarzan, add *cat bridges* to a set of climbing trees.

Get lengths of scrap lumber long enough to span some tall pieces of furniture, kitchen cabinets, and cat trees. Your cat will appreciate the quick, safe access to a lofty personal perch from which she can survey her world. Even the laziest couch-potato cat might be tempted by the high-altitude views and the chance to practice that quintessentially feline *I'm-really-above-it-all* attitude.

Build steps spiraling up a wall, or lay boards from one high surface to another. Your custom-designed aerial cat transportation network will encourage your cat to explore, stretch, and exercise—invaluable for alleviating boredom and preventing unhealthy weight gain. Options for expansion are limited only by your imagination. Sturdy sisal or nylon hammock netting attached to a wall, draped down at an angle, and firmly affixed to a low bench or shelf makes a terrific scrambling and climbing structure—a jungle gym for your little explorer.

A floor-to-ceiling scratching post or climbing tree makes a safe, sturdy access point for your cat's overhead skyways. Connect the skyway only to areas you feel comfortable with your cat accessing. (Keep dangerous and breakable objects out of range.) As you construct your skyway system, view all parts of it through the eyes of a curious cat who has the ability to cover several feet at a leap. Don't inadvertently provide launch pads to areas you'd rather your cat didn't go.

Need some inspiration? Look for *The Cat's House* by Bob Walker and Frances Mooney. In their delightful, colorfully photographed book, this cat-loving couple shows how they've transformed their home into a feline paradise.

SAFETY CHECK

Whatever kind of elevated experience you provide for your cat's exercise and amusement, monitor all structures and their component parts frequently and vigilantly for signs of slipping, damage, or other safety hazards. Your cat is equipped with a handy built-in righting reflex, but she can still be seriously injured if she slips or falls unexpectedly.

Outsmarting Curtain Climbers

- Use spring-tension curtain rods instead of rods nailed or screwed into the wall. When your cat tries to scale the curtains, the rod will fall down, making this a no-fun activity.

- Remove curtains or drapes from their rods and reattach them very loosely with thin thread. When your cat attempts to use them as handy ladders, down they come.

- Curtain climbing is a kitten specialty. Consider removing curtains and drapes until your kitten grows up.

- Hang roll-up window shades instead of curtains or drapes.

- The appeal of curtains and drapes lies in their utility as handy, climbable ladders to reach the desired heights. Make sure your cat has plenty of acceptable climbing opportunities and cat-attractive climbing structures, so she doesn't need to scale drapes to reach the heights.

- Hang vertical blinds instead of curtains or drapes. They're attractive, versatile, non-climbable, and close to indestructible.

Outsmarting Toilet Paper Hobbyists

- Keep the bathroom door closed.

- Install the roll of toilet paper backward, so that it unrolls against the wall instead of forward. This won't discourage all cats, but will slow down some of them.

- Install a cover or guard over the roll. You can make your own with some cardboard and packing tape.

- Balance a small cup of water on top of the roll.

- Balance a soda can filled with pennies on top of the roll.

Outsmarting Bedroom Bounders

Some cats just don't know when to say good night. They wake up ready for action at 12:13 a.m., and 2:00 a.m. (and 4:00 a.m. and 5:16 a.m.), demanding food, play, or just attention. It's only natural. Cats are nocturnal hunters with a distinct crepuscular (dawn and dusk) preference. Their wild cousins are most active around dawn and dusk when their preferred prey is out and about. She's just being a cat, but you need your sleep. What can you do?

- Lock your cat out of your bedroom.

- If she scratches or wails at your door, *ignore her*. To outsmart her, outlast her. (It may be tough.) *Don't* let her in, snuggle or try to console her, or, worst of all, feed her. You'll be reinforcing the behavior you don't want.

- Install her overnight in her hideaway with a radio playing softly. Leave a few toys or treats for her to discover on her own.

- Equip her hideaway luxuriously with a comfy bed, scratching post, climbing tree—and plenty of solo toys.

- Conduct a vigorous interactive play session just before bedtime. Wear her out with plenty of pouncing, running, and leaping.

- Conduct an active training session just before bedtime. Many cats enjoy clicker training.

- Settle her down with a pleasurable grooming or cat massage session at bedtime. (See Resources.)

- Feed her supper just before you retire for the night. She'll be ready for a nice snooze afterward.

- Or . . . just before you retire, hide several favorite treats (either in her hideaway, if that's where she spends the night, or all over the house). Pick creative hiding spots: high atop cabinets, underneath throw rugs, in cat beds, behind furniture. Foraging for her treats will keep her busy and happy for quite a while. Between the active play and foraging, she should be ready for a long nap.

Click! Treat! Good Cat!

Old-fashioned animal training uses aversive conditioning: You correct the animal when she *doesn't* do what you want, or does something you *don't* want. Correcting means punishment: pain, or an unpleasant surprise. It causes stress. The animal has to continually *guess* what she needs to do to avoid the pain or surprise.

Modern behavioral science has unlocked the keys to training based on *positive reinforcement*. With clicker training, there's no punishment, no correcting (and we all know how well punishment works with cats)—just a click (an audible marker signal). The click isn't to get your cat's attention. It means, *You're doing something I want you to do and I'm going to reward you for it*.

Clickers, widely available at pet stores, fit in your palm. Keep one handy, maybe in your pocket or on a cord around your neck. Watch for the behavior you want. The instant you see it, click, and pay off. The reward can be praise, petting, a cat treat—whatever your cat most craves.

See the desired action. Click. Treat. It's that simple.

Cats take especially well to clicker training because there's no downside. It's the "good deal" that opportunistic cats are always looking for. It exercises their minds and shows off their intelligence and adaptability. A clicker-trained cat delights in trying new things, because she knows there's no punishment. What's the worst that can happen? No treat. A clicker-trained cat is happy, confident, joyful—*Hey, this is fun*. Training is play—a game, an everyday bonding ritual.

See the Resources section to find out more about clicker training.

- Be sure she gets enough activity, play, and stimulation throughout the day, every day, so boredom and loneliness don't cause her to seek you out obsessively at night.

- Is she howling at (or to) a cat outside? Nighttime yowling may be caused by the presence of an NC (Nobody's Cat) or OPC (Other People's Cat) wandering and calling outdoors. The intruder's pheromones and voice travel easily indoors. See chapter 13 to learn ways to discourage OPCs and NCs.

- If an outdoor cat is causing restlessness for your indoor cat, close windows and pull drapes to minimize incoming scents and sounds. Play a radio or CD player softly.

- For extreme cases: Set up the vacuum cleaner just outside your bedroom door. Plug it in and have a way to turn it on from your bed (using a remote control device, usually sold for turning lamps on and off remotely). When your cat starts howling and scratching at the door, let the vacuum monster rip. (Okay, this is not "nice." But it's for *extreme* cases only.)

If you do choose to share your bedroom with your cat, be tolerant and under-standing of her sleep patterns. Your cat will likely come and go. She'll share your bed for a few hours, leave on some urgent kitty business, and return later. "Cats go where they please and please where they go." If you'd like to minimize conflicts but still share some furry nighttime warmth and purring, try these tips:

- Keep a few cat toys handy on your nightstand. A laser pointer is great for amusing a restless cat in the middle of the night.

- If your cat wakes you up by pouncing on your toes, stop moving your toes. Toss a small toy across the room.

- Set up an irresistible cat bed in your bedroom, but not on your bed. She may prefer her own bed to your tossing and turning.

If you suffer from allergies, it might be best to ban your cat from your bed-room no matter what you (or she) would prefer. Neither of you will benefit if you're an exhausted, sneezing zombie all day. (See chapter 11.)

Outsmarting Armpit Nuzzlers

Many affectionate cats make their owners puzzled and uncomfortable by their insistence on sniffing and snuggling armpits. Are they being . . . weird? No. It's normal, natural, and should please you immensely. If your cat does this, she's finding a pleasurable outlet for natural caretaking and scent-mingling urges that, in kittenhood, were directed toward MomCat and her littermates. She's enjoy-ing—actually "smell-tasting," with her vomeronasal organ as well as her nose—your familiar, beloved odor and merging it with her own, adding it to the group scent. She's welcoming you back to the group, welcoming you home. Congratula-tions. You're an honorary cat!

She's also trying to figure out where you've been and what you've been up to. She's a cat, not a person, so her investigatory efforts are scent-based rather than verbal. She can't ask, *Where did you go today?* or *Did you meet any strange kitties?*

But she can tell by the smell—just as she'd do with another cat. She's transferring a normal, typical cat behavior into a human context, where it might seem weird. But enlightened owners enjoy it.

Still uncomfortable? People often are, especially when their cat's behavior is focused on odors that humans find unpleasant or socially undesirable. Be patient, and let her sniff. Most likely, she'll be satisfied with a brief, intense interaction. If she keeps it up longer than you like, though:

- Divert her attention.

- Stand up and move around.

- Change positions or move to a different chair. This can help break the pattern or context in which the unwanted behavior started.

- In a gentle, friendly way, make her target area less accessible.

- Get up and start an interactive play session, or serve dinner.

- Don't push your cat away. It's rude and insensitive, and even negative attention may reinforce the behavior.

This intimate, scent-oriented sharing is a highly motivated behavior—and a sign of trust and love.

It's always easier to prevent the onset of an unwanted pattern than to try to extinguish an established routine. By carefully observing your cat's habits and predilections, you'll learn to recognize when she's about to engage in an activity you want to discourage. If flopping down in your reclining chair at the end of the day signals your cat to begin a dedicated armpit-nuzzling session, break up your routine. Go to another room, sit in a different chair.

Or divert your cat's behavior to another object, a scented stand-in. Rub a small, soft cat toy on your armpits, or provide the cat with an unlaundered sweatshirt or other article of your clothing. Your cat might appreciate access to your shoes, or dirty socks. When your cat's behavior becomes a bit too much for you, refocus her toward the substitute.

Do you wear a perfume, aftershave, or other scent your cat finds particularly attractive? (Wildcats have been known to react to certain perfumes and cleaning products as if they were premium catnip.) Alter your scent signature to one that's less arousing to your cat.

Because scent is so meaningful to cats, and because they're so much more sensitive to odors than we are, we often forget that scents barely noticeable to us can carry powerful messages. Some soaps, lotions, beauty products, colognes, and deodorants partially mimic feline pheromones, causing cats to react according to deeply held natural mating or maternal behavior patterns. If you have a particularly persistent armpit nuzzler or neck licker, try changing product brands or minimizing your use of scented products.

Your cat loves routine, ritual, pattern. Work on establishing new patterns and rituals that make both you and your cat happy and comfortable with each other. The stress and conflict caused by persistence of a behavior that makes you uncomfortable or upset may, over time, erode the bond between you and your cat, or keep that close, loving bond from developing in the first place.

Outsmarting Keyboard Tappers

If your cat shares your home office, it's likely that she'll help by walking on your keyboard, with interesting results, up to and including deleting documents and crashing your computer.

If your cat has learned how to log onto the Internet, surf cat shops, and buy kitty treats with your credit card, your problems are probably beyond the scope of *Outsmarting Cats*. But if it's still just random paw tapping, you can outsmart her. Here are some ways to outsmart your keyboarding kitty:

- Keep your keyboard and mouse on a sliding tray or drawer that retracts under your desk, monitor, or work surface. Whenever you leave your desk, slide the tray in.

- If you don't have space for a sliding tray or drawer, or if your desk design makes such a contrivance awkward, get a keyboard cover, a rigid plastic box that expands to cover both keyboard and mouse. Put the cover in place whenever you leave your desk. You could probably cobble one together yourself. Or use a long rectangular plastic storage dish.

- If the hardware approach seems too mundane, try a software solution. A program called PawSense (see Resources) detects "cat-like typing." Depending upon how you set up the software, it sounds one of a variety of loud, cat-aversive alarms through your computer's speakers. It will also refuse input that it decides is coming from your cat (cat-like typing) rather than from you.

Outsmarting Monitor Perchers

Cats love to perch atop old-style CRT computer monitors: relatively flat, warm, and near their favorite people. Some people don't mind. (Be sure to vacuum the cat hair out of the top vents occasionally.) If monitor perching offends you, or if those big kitty eyes following every keystroke gives you the heebie-jeebies:

Chrysanthemum tries yet again to nab that maddeningly elusive screen-mouse.

- Offer your cat a heated pad in her climbing tree, or a heated kitty hammock by a window. She may decide that it's more comfy than your monitor.

- If possible, set up the heated pad in close proximity to your computer. It's likely that your presence accounts for some of the monitor's cat appeal.

- Try a kitty monitor shelf, a padded board that attaches securely to the top of your monitor. It offers your cat a soft, padded perch that also catches cat hair before it wafts into the monitor's vents. (See Resources.)

- Treat yourself to a flat-panel monitor. There's precious little room atop these skinny devices for even a slim cat to perch. Plus, they don't exude the comfy warmth of the old-fashioned tubes. (Never underestimate a determined cat, though. She may do it just to prove it can be done.)

- Make your home office a cat-free zone.

Outsmarting Kittenish Claw Scramblers

Kittenhood is fleeting, but kitten damage can last forever. Kittens under six months are still unsure of their strength and leaping abilities, and often choose to scramble rather than leap up to where they want to be. Kittens scramble with all claws fully extended because they feel more in control. Needless to say, this is not healthy for most furniture.

Here's how to outsmart kittenish scramblers:

- Keep your kitten's claws clipped. This helps prevent damage, and accustoms your kitten to regular claw maintenance—a great idea if you plan to keep her claws clipped when she's an adult. (See chapter 6 for more on claws and claw clipping.)

- Drape wooden tabletops with heavy blankets (Vellux and similar thick blankets). Don't place large or heavy objects on draped tabletops. Kitten could pull them over.

- Fit wooden tabletops with heavy tempered glass tops with rounded, smoothly finished edges. These are relatively inexpensive; have them custom-cut locally.

- Place particularly valuable or delicate wood furniture in a cat-free zone until your kitten is regularly leaping rather than claw scrambling.

- Beware of tippy, top-heavy tables with only a single central supporting post. These are too easy for a leaping kitten or cat to tip over. Look for sturdy, four-legged tables.

Let Your Cats Help You Outsmart Potential Burglars

You can take advantage of your cat's predilection for jumping up on lamp tables to help protect your home. How?

Law enforcement personnel advise homeowners to equip their homes with automatic lights that go on at sundown and off at sunrise, the better to throw off potential burglars and intruders, especially while you're away from home. This is good advice, but if you have cats, there's a better way: a system for which no burglar will ever figure out the timing or the logic. Because it isn't human logic at all—it's cat logic.

Equip a number of ordinary table lamps with motion-detecting switches. Whenever a cat comes in the vicinity, the light comes on. You can set these devices to keep the light on for as brief a time as a minute or up to fifteen minutes. Combined with one or two off-at-dawn, on-at-dusk lamps, your cat-controlled lighting patterns will mystify and confuse even the most intrepid cat burglar.

- Don't put valuable or breakable lamps or other items atop small or low tables. Use sturdy furniture with four legs or a solid base.

- Secure small breakables on shelves and tables with dabs of sticky putty or gel (The Museum Putty and The Museum Gel; see Resources) or small hook-and-loop (Velcro) dots. These materials can prevent accidental knock-overs, but they won't perform miracles. If your cat or kitten is determined to push something over, she will.

- Especially while your kitten is young and klutzy, don't place dangling runners, tablecloths, or other linens atop furniture. A claw can catch an edge, sending cat, table, and everything on the table tumbling.

- If you rely on placing breakables and valuables on high shelves to protect them from leaping cats, be *absolutely* sure that your cat can't get up that high. Your cat can leap several feet up from a standing start. Make sure there are no nearby pieces of furniture, window ledges, or other launch pads from which she can mount an aerial assault on the forbidden heights. Use your imagination—your cat will. Be especially aware of potential launch pads that are in your home temporarily (ladders for a repair or construction project) or seasonally (your Christmas tree).

- When arranging books and other objects on your bookshelves, line them up as close to the front edge of each shelf as possible, to deny your cat attractive landing spots on the front edges of the shelves.

- In your decorating and furnishing, avoid putting temptations (attractive nuisances) in your cat's or kitten's way. It's not fair to her, or to you.

- No matter how careful you are, be prepared for some damage to your furniture. Keep some patch and refinishing supplies on hand for quick touch-ups. (A Tibet Almond Stick is a must for cat owners; see Resources.)

Chapter 13

OPCs (Other People's Cats) and NCs (Nobody's Cats)

Indoor, Outdoor

Here's the good news: According to the 2011–12 APPMA (American Pet Products Manufacturers Association) National Pet Owners Survey, 60 percent of the 86.4 million owned cats in the United States enjoy an indoor-only lifestyle, and an additional 10 percent of cats are kept indoors at least at night. And many owners of indoor-outdoor cats are planning to make their next cat an indoor-only cat. This reflects the growing American appreciation of cats as full members of the family, entitled to all the benefits of a safe, enriched indoor life. This is good news for *all* cats, and for birds and other wildlife, gardeners, kids with sandboxes, and exhausted humans kept up until all hours by caterwauling, battling, sex-crazed felines.

Whose Cat Is That?

When you see an unknown cat in your neighborhood, your first thought is that it's someone's cat, out for a stroll. It may be. But it may also be a lost, stray, or feral cat. It may be ill, or in need of assistance.

- Don't assume all wandering cats are strays or ferals. Many owners believe in giving their cats freedom of the streets. If the cat isn't causing trouble, let him go on his way.

- Try to determine if it's an owned cat. If he runs away at your first approach, or is extremely wary and unfriendly, it's likely a feral cat, or a long-term stray (once-owned) cat.

- Does he have a collar? Keep a pair of binoculars handy to observe the cat from a distance. If he's approachable and has a collar, check it for phone number or other identification.

Feral? Stray? Or a pet out for a stroll?

- Is he injured? Limping? Do you see wounds? If so, call a qualified animal control officer or veterinarian for advice. Don't approach an injured or ill cat. He may be in pain and lash out.

- Is it a very young kitten? MomCat and the rest of the litter may be nearby, and may be in danger. Call a local cat rescue group or veterinarian for advice.

- Is the cat in imminent peril? Are kids tormenting or harassing him? Is severe weather expected? Call law enforcement, animal control officials, or a local cat rescue group at once.

- Does the cat come around regularly, but still seem to be homeless? Ask local veterinarians, shelters, local officials, and police departments if anyone has reported a lost cat. Describe the cat in detail. Note color, markings, eye color, fur length, size, and gender. If possible, use a zoom lens and take a photograph. Write down your observations, and the date, time, and location.

- Many owned cats carry *microchip identification* (an electronic chip underneath their skin, usually on the back between the shoulder blades). A local veterinarian or shelter might have scanning equipment to check if the cat is chipped. If he is, you can read his ID and get him safely home.

- Does the cat have a tipped ear? (Guardians of managed feral cat colonies mark each colony cat by cleanly nipping off a tiny bit of ear tip when the cat is spayed or neutered.) He might belong to a local colony, or be from a feral group that's no longer actively managed. Call Alley Cat Allies, Alley Cat Rescue, or a local feral cat group for advice. (See Resources.)

- If you can't locate an owner, refer the cat to a local shelter, cat rescue group, or local animal control officials. *You're not alone. There are lots of people ready and willing to help.* Start with Pets911 (888-PETS-911, or www.1888PETS911.org).

- If you decide to adopt the cat yourself (if he's truly homeless), be sure he's free of contagious diseases before introducing him to your other cats, other pets, and family. Be sure he's altered, too, or schedule the surgery.

- Know the laws in your area. In some places, a stray cat must be taken to an approved shelter to give the cat's owner a chance to find him. If you want to adopt the cat if no one claims him, make sure the shelter knows this. Check back frequently.

- Don't transport and release a troublesome stray in a different area, however distant. It's not fair to the cat, or to the residents there (human and otherwise).

- If you can't keep him yourself, *don't* place a "free to good home" ad. Not everyone who answers such ads has the cat's best interests at heart. Work through a reputable local shelter, rescue group, veterinarian, or people you know well to find the cat a secure, permanent home.

Approaching a Cat You Don't Know

It's wise to think of *all* unknown cats as wild animals. Even once-owned cats who've become lost, run away, or been abandoned quickly revert to wild, sometimes unpredictable, behavior. They're also subject to numerous diseases, some of which are *zoonoses*, diseases that can be transmitted from animals to humans. Ill, injured, or abuse-wary strays and ferals can be aggressive. Worse, they can seem tame and friendly, only to lash out if frightened, cornered, or startled.

Courtesy, caution, common sense, and respect are essential in approaching and handling *any* cat, even your own. They're doubly important in approaching a cat you don't know, especially one who may be poorly socialized, fearful, or unaccustomed to humans.

- Never assume *anything* about any cat you don't know.

- If the cat's owner is known and nearby, ask permission before touching or handling the cat, or offering food. (Some cat owners, especially those with pedigreed cats at cat shows, don't let anyone touch their cats for fear of possible disease transmission.) Respect the owner's wishes.

- Speak soothingly and softly, in a relatively high, kittenish pitch, which sounds friendlier and more reassuring to cats.

- Avoid harsh or sudden sounds (a sneeze, blowing your nose), especially anything like a hiss or growl.

- Stay in one place, or move slowly. Approach from the cat's front, but slightly sideways. If you walk rapidly or straight toward the cat, he may feel threatened and run. Keep your hands visible. Don't trap or corner him.

- Don't hover over the cat.

- When you're about ten to twenty feet away, crouch slowly, to ground level. Keep speaking softly.

- If the cat still seems comfortable, slowly extend one hand. Let him approach you at his own pace. He'll probably sniff your hand.

- Place small bits of tasty food in your open palm, and on the ground in front of you. Let him make the next move.

- Don't try to pick him up. If he's particularly friendly and trusting (an OPC rather than an NC), he may crawl onto your lap. Feel free to pet. Don't restrain him on your lap or pick him up.

- Let him leave if and when he wants to.

- If he's an NC (stray or feral), or an OPC whose owner you don't know, be cautious. Most stray and feral cats flee quickly. But if the cat lets you approach or even touch him, stay tuned to his mood and body language. Feral and stray cats are much warier than owned cats, and may lash out if cornered or threatened.

- Leave him an obvious, easy escape route. If he's injured or obviously ill, and you're trying to capture him for veterinary care, it's safer to use a humane box trap than to try to pick him up yourself. (Before attempting to trap, check local laws. Trapping animals without a license is against the law in some areas.)

In from the Cold

You've decided to adopt, and bring indoors, a stray cat you've been feeding for several months. Or maybe the nighttime howling of coyotes has motivated you to treat your indoor–outdoor cat to a safer lifestyle. You need to persuade him that indoor life is a much sweeter deal than roaming the neighborhood, or 24/7 access to the kitty door.

You can't trick him, or convert him against his will. But you can outsmart him. Redirect his attention, reorient his focus to his exciting new lifestyle. Highlight, in ways he'll understand and appreciate, the benefits and joys of life indoors. Make it worth his while to go along. Pour on the playtime and pampering. He'll soon transfer his focus from his old outdoor haunts and pursuits to his deepening relationship with you and your family.

Some cats take to indoor life quickly and easily. Many behaviorists recommend the all-at-once approach—getting the worst of the transition over as soon as possible. If you go cold turkey, remember that even minor backsliding will make the whole job much tougher. Be consistent. If you declare outdoors off limits, and your cat escapes even once, he's much less likely to take your determination seriously. The race-for-the-door game gets old, fast.

A few cats, especially old wanderers who've spent years outdoors, seem to have a lot of trouble adjusting. If your cat starts spraying, house soiling, endlessly scratching at the door, howling, being uncommonly aggressive, or refusing to eat, a trip to your veterinarian is in order. A short course of certain types of mild tranquilizers or other drugs can help get "problem adjusters" over the hump.

Don't try to diagnose and medicate your cat yourself, though. Your veterinarian has the necessary skill and experience to evaluate the situation and prescribe the best medication, dose, and duration of treatment. And no drug is a magic "quick fix." Your cat will still need your understanding, support, and patience, and you'll still need to enhance his new environment to maximize his comfort and sense of well-being.

If you've handled a stray, feral, or other unknown cat, don't put your indoor cats (and family) at risk for zoonoses and other contagious diseases. Wash all exposed skin (especially your hands) with antibacterial soap and change your clothes before going near your own cats or kids. Never let your cats near any carriers, blankets, food or water bowls, or other items that have been used or touched by cats with unknown health status.

Outsmarting Troublemaking OPCs and NCs

Other People's Cats (OPCs) and Nobody's Cats (NCs: ferals and strays) are notorious for making nuisances of themselves. They bask in the sun on car hoods (leaving muddy kitty footprints on the way), dig in garbage cans and scatter trash, use flower gardens and kids' sandboxes as litter boxes, chase birds away from feeders, stalk and kill birds and small animals, and scratch kids who try to clutch and pet them. Punishing a cat, your own or an OPC or NC, for these infractions is just as useless as punishing your indoor cat for urinating on the carpet or scratching the sofa. He won't know what you're upset about—and it won't work.

IS THAT YOUR CAT?

If the offending feline is your own, consider bringing him indoors to stay. If the cat belongs to someone you know, speak frankly with them about the trouble their cat is causing. They may respect your complaints and make a sincere effort to control the problem. Offer as much help as they'll accept. Introduce them to your own veterinarian. Offer to loan them *Outsmarting Cats,* or buy them their own copy. Be a resource.

If the troublemaker is an NC or an OPC, the dangers posed by zoonoses, bites, and other health risks, as well as property damage and lessening of your family's quality of life, sometimes justify using aversive stimuli, repellents, and deterrents.

Whenever using aversives, repellents, and deterrents, remember that the cat you're dealing with may be someone's beloved pet. But no matter the cat's origin or ownership status, stick to safe, humane approaches, no matter how frustrated you are. Especially avoid poison. It's inhumane, usually illegal, and puts other pets, wildlife, and humans at risk.

OUTSMARTING YARD AND GARDEN INTRUDERS

That enticing expanse of fluffy, freshly turned earth might look like a newly planted seedbed to you, but to a cat it's an ideal open-air toilet, or just a great spot to dig, scratch, and take a nice dirt bath. Here are some proven stratagems for outsmarting garden-marauding cats:

1. Most feline garden mischief is directed at soil, not foliage. Cats seldom bother mature plantings. It's dirt they're after. Make the dirt either inaccessible, or difficult or unpleasant to dig in.

2. Lay light, twiggy brush, or thorny cuttings from rosebushes or berry bushes, across seedbeds. This discourages most digging, but allows light and air to reach plants.

3. Sprinkle pinecones, or pinecone pieces, thickly in garden beds.

4. Lay one-inch chicken wire across seedbeds. Weigh it down with rocks or bricks. Cats can't dig through it, and wire irritates their paws. Plants easily emerge through the openings, and you can leave the wire in place throughout the growing season.

5. Poke twigs and small branches, or inexpensive craft sticks ("Popsicle sticks"), available at craft stores, vertically into the soil throughout planted flower or veggie beds, especially where cats have been digging.

6. Surround ornamental plantings with a layer of river rocks or pebbles.

7. Try cat-repellent sprays, homemade and commercial. Mixtures containing garlic or citrus oil deter some cats. Reapply at least once a week. Here's one recipe: Pulverize two to three cloves of fresh garlic and about four hot peppers in a blender. Add this goop to a gallon of water. Add a few drops of liquid dish detergent (helps the mixture stick to surfaces). Sprinkle around the area intruders frequent.

8. For another homemade cat deterrent, save dried used coffee grounds and crushed dry eggshells. Combine, and spread on a cookie sheet. Spray generously with cooking oil. Sprinkle with powdered cayenne pepper, mix, and sprinkle again. Every few weeks, distribute small amounts in your flower beds, or on the path intruding cats usually take. It's long-lasting, biodegradable, and safe. Sprinkle this mixture in different, unpredictable places each time, and invaders will soon learn to avoid your garden.

9. Fill small cheesecloth bags with mothballs or moth crystals, and sprinkle them around the beds. Cats dislike the smell. You'll need to add more after heavy dew, rain, or watering, as they melt quickly when wet. (Never use mothballs as cat deterrents in enclosed areas—the concentrated fumes can kill cats in a short time.)

10. Sprinkle hot red pepper flakes over the beds. Visiting cats will get a spicy surprise when they lick the pepper off their feet.

11. Spread chicken manure around your garden. (Check local egg farms for inexpensive sources.) Cats dislike the smell, and the manure is good for plants.

12. Cats dislike digging in wet soil. Wet garden soil in the evening, because many OPCs and NCs like to visit at night. (Don't wet foliage, just the ground.) Keep this up until the intruders get used to going somewhere else.

13. Physical exclusion—using barriers—is better than any repellent. Old-time gardeners swore by dense hedges of common greenbrier (its antique name was cat-brier) to keep cats out of kitchen gardens. Greenbrier is also a favorite of birds, according to an 1859 horticultural reference.

14. Use cardboard or plastic mulch, or a mulch made of very large, rough bark chips. (Cats, unfortunately, adore digging in soft organic mulches.)

15. Fence in your garden with chicken wire or chain-link fencing. Keep in mind how high cats can leap. (A sturdy fence will also keep out many other garden marauders such as deer, raccoons, rabbits, and dogs.)

16. Use bird netting. Cats get their feet tangled in it and will avoid areas where it's been laid over plants.

17. If cats are walking along a wall or fence to get into your garden, apply a sticky substance (e.g., Tanglefoot) along the fence top. Cats hate sticky stuff on their paws.

18. Balance trays of pebbles on fence tops, where cats will knock them over noisily when jumping up. To outsmart car-hood loungers, balance trays of pebbles at the edges of the car's hood and roof.

19. This one sounds weird, but it sometimes works. Save the solids (the feces) from your indoor litter boxes, and distribute them around the perimeter of your property. Wildcats often mark their territories with scat piles, visual and scent marks of ownership.

20. Mark your property borders with citrus peels, or small rags or sponges soaked in ammonia. Refresh these at least weekly as the odors fade, until you're sure the intruder has stayed away for at least two weeks.

21. If you catch the intruder on your property, startle him with a sudden noise. Fill a soda can with pebbles or coins and keep it handy on the porch or in the garden for rattling.

22. Get a motion-detecting device that turns on a sprinkler or bright light in response to motion in your yard or garden. Few cats will relish being suddenly doused when they think the coast is clear.

Farmer's Helpers?

Cats in your garden aren't always bad news. If you're trying to grow root crops like carrots, parsnips, or beets, a regular cat patrol can persuade crop-destroying moles, voles, gophers, and other tunneling underground pests to take up residence in less stressful gardens.

23. If an OPC or NC has sprayed or marked your front porch or around your house (which happens more often than you might think), wash down the areas with a garden hose and follow up with a dousing with plain white vinegar, or use an enzymatic cleaner or pheromone extractor. This will help keep the intruder from returning again and again to re-mark "his" territory, and also help keep his scent from wafting indoors and upsetting indoor cats.

24. To keep NCs and OPCs away from your porch or deck, spritz the area with a mosquito repellent (one made for use by humans) containing oil of citronella and other strong-smelling plant oils. Cats dislike the strong odors, especially citronella. (Don't use citronella-scented lamp oil or torch oil—these are flammable!)

25. To keep cats off garden and yard structures (decks, lawn furniture, cars, fences), use the same strategies as for keeping indoor scratchers away from sofas. (See chapter 6.)

26. Plant aromatic herbs and flowers such as rue, marigolds, allium (onion family plants), or chamomile. Most cats find these distasteful.

27. Plant the newest sensation from Europe, *Coleus canina* x *Plectranthus* hybrid. It's common names are Pee Off (or Piss Off) and Scaredy Cat. Hybridized in Germany, it's now available in the United States. (See Resources.) It's an annual foliage plant that grows twelve to twenty inches tall, can grow in most soils, and produces attractive blue flowers. It releases a strong odor that's said to be repugnant to cats (and foxes, dogs, and martens) but is undetectable to humans unless the leaves are crushed and rubbed between the fingers.

28. Outside your garden fence or on the edge of your property, build a sand trap, framed with scrap lumber and filled with sand. OPCs and NCs will likely ignore your garden and make a beeline for their own sandbox. A plain pile of sand may do as well.

29. Establish a diversionary catnip patch at a distance from your garden. Set a few sturdy *Nepeta cataria* plants in a sunny, well-drained spot. The local felines may spend so much time enjoying your herbal hospitality that they'll forget to dig in your flower beds.

Whose cat is that?

OUTSMARTING SANDBOX DIGGERS

To a wandering cat, a kid's sandbox is a handy litter box. Outsmart this bounder by:

- Covering the sandbox with a sturdy cover or large sheet of plywood weighted down with bricks whenever it's not in use.

- Offer an alternative: a pile of sand or dirt well away from your kids' play area. Plant some catnip plants nearby.

- Use one of the other cat-deterrent methods described above.

OUTSMARTING BIRD STALKERS

Want to start a really big fight with your neighbors? Bring up the subject of cats vs. birds. Do cats really murder billions of helpless birds every year? Are cats responsible for plummeting songbird populations? Few subjects cause more animosity, bickering, and controversy. Even scientists and researchers can't resist trading barbs and dueling with data sets over this still-unresolved issue.

Studies of cat predation, and a commonsense look at cats' hunting preferences and techniques, show that cats have an overwhelming preference (70 percent or more) for ground-based prey (chipmunks, rabbits, rodents, shrews, voles, and so forth) and soft-bodied insects. Opportunistic hunters, cats nab whatever's most avail-

Young Tiger discovers to his chagrin that nabbing chickadees is not as easy as it looks.

able and easiest to catch. Overeager kittens jump at birds, but cats learn early that bird hunting is hard work (birds can fly), and not particularly rewarding.

One study found that the major predator of birds at feeders was not cats, but sharp-shinned hawks. Songbird populations are declining in many areas, but cats are likely *not* a big factor. Habitat loss and fragmentation, road incursions, development, pollution, collisions with plate-glass windows and towers, and exhaustion and confusion caused by bright lights at night kill many more birds than cats.

Still, cats *will* stalk and kill birds—if they can get them. If you set up bird feeders near an ideal stalking lair, or where cats can easily sneak up on ground-feeding birds, you're going to lose a few feathered friends.

The mere presence of a patrolling cat (your own cat, or an OPC or NC) on your property can discourage birds from living, nesting, and feeding there. It's not that cats have killed all the birds; they've just flown off to less stressful quarters. To outsmart furry bird stalkers:

- Fence your yard, and top your wall or fence with an inward- or outward-slanted shield (made of wire or fencing, such as Cat Fence-In) to keep OPCs and NCs out, and your own cats safely in.

- Set up bird feeders in open areas, well away from brush, bushes, decks, and other cover cats can use to stalk birds.

- Mix a little Tabasco sauce or cayenne pepper powder in with the seed in your feeders. It doesn't bother or harm the birds (they can't taste it), but it does irritate mammals and rodents when they walk in it.

OUTSMARTING THE MIDNIGHT CHORUS

That loud, endless, nocturnal wailing, screeching, yowling, and meowing is almost always about . . . sex. Females in heat call enticingly to males, and hordes of rival males argue, bicker, and fight over access to them. Neutering and spaying the offenders will solve most of your noise problems. If the singers are OPCs, and you know their owners, initiate a frank dialogue about the problem, and gently suggest the solution.

OUTSMARTING TRASH SCATTERERS

Cats invading your garage to sample the trash, or scattering rubbish left out for the trash collectors?

- Sprinkle garbage bags and cans and the ground around them with curry powder. (Buy it cheaply in bulk at Asian supermarkets.)

- Use just-in-time trash management: Put out receptacles just before the trash collector is due.

- Splash trash bags and garbage cans with white vinegar, ammonia, or other nonpoisonous, sharp-smelling liquid.

OUTSMARTING TERRIBLE TOM, NEIGHBORHOOD TERROR

Sometimes one cat can cause headaches for a whole neighborhood. These serial offenders are usually sexually intact males, but some turn out to be altered (and currently owned) cats.

The troublemaker wanders the neighborhood looking for trouble and leaving stinky calling cards in gardens and sandboxes. He yowls and caterwauls at all hours. If he's not neutered, he pursues the slightest hint of females in heat, and noisily challenges other males to battle. He sprays and marks the windows, doors, and porches of indoor cats, who spray indoors in retaliation. He scratches and bites kids (and grown-ups) who try to pet or play with him. He chases and harasses other cats and dogs, scatters birds from feeders, invades garages, and rips up trash bags. In other words, he behaves like a rude teenager on spring break.

Terrible Tom, the neighborhood terror

He's also the reason a lot of otherwise nice folks turn to poison, inhumane traps, and even shotguns.

If the offending cat has owners:

- Talk to the owners, tactfully, about the cat's disturbing behavior.

- Mention local noise control and animal control ordinances. (Do your research first.)

- Offer information on spaying and neutering, especially free and low-cost clinics. Tell them about discounts available in connection with Spay Day USA (the last Tuesday in February, every year).

- If money seems to be an issue, offer to pay for spaying or neutering yourself, or take up a neighborhood collection.

- Offer information about the dangers of unsupervised outdoor life for cats.

- Offer information about the pleasures of a safe, indoor-only lifestyle for cats. Use your own indoor cats as happy examples.

- Don't nag. Be friendly, positive, upbeat, and cheerful. Your goal is to help them and their cat, as well as trying to get some peace and a little sleep.

- Lend them *Outsmarting Cats,* or buy them a copy.

- If all your sincere efforts at diplomacy fail, notify animal control officers about the continuing problem.

- Meanwhile, play a radio or favorite CD softly at night to mask all that cater-wauling, and tell your kids to stay well clear of the offender.

Trapping the Offender

If the cat is stray or unowned, you may decide to trap the offender and turn him over to a reputable shelter or cat rescue group, or to a new owner who agrees to give him a secure, permanent home. Ensure that whoever takes responsibility for the cat will have him neutered and fully vaccinated.

- Check local ordinances first. Trapping animals without a license is against the law in some areas.

- Seek advice from an experienced person or group, such as a shelter or cat rescue group, before proceeding, especially if you haven't trapped a cat before. (See Resources.)

- Formulate a well-thought-out plan for what you'll do with the cat once he's trapped.

- Obtain a sufficiently large humane box trap. (Borrow one from your local animal control officer, shelter, or rescue group.)

- Time your trapping. If you plan to take the cat to a shelter, don't set the trap if the shelter will be closed the next day.

- Closely supervise and monitor the trap.

- Watch the weather. Never expose a trapped cat to extreme heat, cold, wind, or rain.

- Use common sense. A trapped cat is a highly stressed animal, and should be considered dangerous.

- After trapping, cover the trap with a blanket to lessen the trapped cat's fright and stress.

- Keep the cat trapped for as short a time as possible.

OUTSMARTING COLONY WOES

A group of stray or feral cats, called a colony, can become a menace to health and safety for its neighbors of all species. A large colony of unaltered cats will reproduce continuously and rapidly, producing litter after litter of kittens. Unmanaged and unmonitored colonies can harbor reservoirs of parasites, zoonoses, and diseases that threaten humans (especially children), owned cats, other pets, and local wildlife. A colony may have set up cat-keeping in a dangerous or inappropriate location, or where their presence is unwanted. Unfortunately, the traditional municipal and law enforcement response to the presence of feral cat colonies is to kill all the cats, as soon as possible and by any means available (often inhumane).

But the presence of a colony in the first place means that the location offers enough resources to support a group of cats. If the colony's cats are killed or otherwise removed, but the resources remain (as is usually the case), a new group of ferals soon appears to use those resources. They reproduce rapidly. Wandering newcomers join. Soon there's a new colony.

If a large number of unowned cats regularly puts on midnight concerts in your neighborhood, invades sandboxes and gardens, scatters trash, and stalks bird feeders, your neighborhood may be unwillingly hosting an unmanaged feral colony.

A Helping Hand, and More

Humane, responsible help *is* available. National organizations and many small groups and individuals all over the country stand ready to help find humane solutions to your problems with Territorial Tom, Spraying Sam, and Ever-Pregnant Homeless Mama and their many friends and innumerable litters of kittens. Don't go it alone. Contact one of these feral cat management groups for advice.

TNR: The Humane Solution That Works

It's devilishly hard to pin down the number of unowned cats in the United States; estimates range from fewer than the number of owned cats to twice that number. (The higher estimates generally come from bird and wildlife organizations.) To complicate matters, there are two kinds of unowned cats: feral and stray. The same cat can fall into both groups at different times in his life.

Feral cats are born to feral mothers and are completely unsocialized to humans. In fact, most feral cats can never be socialized to humans enough to live

Turtle Comes Home

People who devote their lives to rescuing feral cats are quick to point out that feral kittens, if rescued early enough, can be tamed and socialized and become wonderful family companions. But they also know how difficult it is and it's usually impossible to find permanent adoptive homes for adult ferals. Most are untamable, even unhandleable. Even providing basic veterinary care for them requires a brave heart and heavy gloves.

But some cats have an uncanny way of being the exception to the rule. In *Maverick Cats,* her landmark study of the lives and natures of feral cats, Ellen Perry Berkeley tells the curious story of Turtle. One day this completely feral tortoiseshell cat, born and raised in the cold, unforgiving New England woods, wandered warily onto Berkeley's country property. After many adventures (including bearing kittens) and years of feline hesitancy and indecision, Turtle finally decided to come in from the cold. She lived out the remainder of her years as if she'd always been a pampered housecat, serving as muse and daily inspiration to Berkeley and her writer husband.

Though Turtle was an exception, her transformation from wary feral to settled domestic cat shows that even a fully mature, adult feral cat can become a successfully socialized pet. Mostly, though, Turtle's tale proves that we still have a lot to learn about what cats want and need, and what's important to them and why.

as pets. Very young feral kittens *can* be socialized if they get human attention and handling from the beginning.

Stray cats are (or once were) socialized to humans but have no homes. Most strays can be resocialized and can live out their lives with human families.

According to a 2003 paper published in the *Journal of the American Veterinary Medical Association* ("Evaluation of the Effect of a Long-term Trap-Neuter-Return and Adoption Program on a Free-roaming Cat Population"), "The free-roaming cat population consists of both socialized stray cats and un-socialized feral cats. Individual cats may have a variety of lifestyles during their lives, including owned pet, stray, and feral status. Because of the vast overlap of lifestyle and socialization status continuums, it is difficult to define discrete populations of free-roaming cats."

Even more difficult than trying to count unowned cats has been the dilemma of what to do about them. The old method was "round 'em up and kill 'em," and many areas of the world still use this tragically inhumane approach. But, as has been found with other "nuisance animals," such as coyotes, snakes, and Canada geese, it just doesn't work. If there's an ecological niche—a food source—for free-roaming cats, then no matter how many cats are killed, more cats will show up to fill that niche. Cats are America's number-one pet, so we are increasingly loath to support lethal cat control methods. Rounding up and killing unowned cats seems wrong because it *is* wrong. And it doesn't work.

As recently as 1990, many municipalities (and other sites where groups of unowned cats gathered in groups called "feral colonies") felt that their only option was to "trap and kill." That all changed with the founding of Alley Cat Allies. Louise Holton, a native of South Africa, and her colleagues had a vision of a more humane and more successful way of managing, and gradually reducing, the populations of free-roaming feral and stray cats. Thus, the TNR movement was born.

The TNR Approach

TNR stands for "Trap, Neuter, Release" and it works. Its success all over the United States and elsewhere has been documented in a growing number of scientific studies. TNR is the only strategy that has been successful in humanely reducing stray and feral cat populations. TNR is now endorsed by major humane organizations, cat registries, and wildlife management professionals.

The TNR approach to feral cat management is a proven success in locations all over the world. Feral colonies generally form around *human-created* food and shelter sources: restaurant Dumpsters, college campuses, military bases, urban alleys, parks, suburban neighborhoods where residents leave food out, etc. The old approach to management was barbaric indeed: The cats were just rounded up and killed.

Holton pioneered TNR first in her native South Africa, and in the United States after she moved here. Holton literally wrote the book on TNR: *Feral Cat Colony Management and Control: Facts and Myths About Feral Cats,* a practical handbook packed with information on feral cats, the history of cats and humans, zoonotic diseases, cat predation, TNR how-to's and much more. (See Resources.)

Holton is the founder of both Alley Cat Allies and Alley Cat Rescue, national organizations devoted to educating the public about feral cats and doing the actual work of rescuing and managing feral cat colonies around the country. Both groups maintain websites (see Resources) packed with information and resources for anyone who is faced with a group of unmanaged feral or stray cats, anyone who feeds a few strays but wants to do more, and anyone who just wants to know more about these remarkable, adaptable brothers and sisters of our cherished house pets.

TNR's message surprises many people: Feral cats are not homeless cats. They have a home: outdoors.

As you might imagine, this is a controversial, even abhorrent position in certain quarters. Bird lovers often blame free-roaming feral and stray cats for the decline in songbird populations. But, as Holton proves in the vast amount of predation data she has collected and analyzed, songbird population decline is chiefly a result of *human* activity, not cat predation.

A "feral colony" of animals doesn't just form for no reason. Any unusually concentrated group of animals forms around a consistent, unusually concentrated source of food. No food source, no animals. If you remove the food source, the animals will either starve or disperse. If you remove the animals, other animals will show up to utilize that food resource.

If the food source remains, though—and most of the human-created ones can't practically be removed—someone is going to utilize it. Raccoons (major carriers of rabies), coyotes, hordes of rodents, or another group of cats will soon appear. So even if you remove the ferals, other animals will show up at the food source.

Enter TNR—the solution that *does* work. Here's how:

1. A group of feral cats is identified around a food source.

2. A small group of individuals—the caretakers—volunteers to manage the colony over the long haul.

3. Appointments and arrangements are made for spay and neuter surgeries, vaccinations, and other necessary veterinary care.

4. The cats are caught in humane traps and transported to the veterinarian.

5. Each cat is examined, vaccinated against rabies and feline viruses, and spayed or neutered.

6. While under anesthesia, each cat is "eartipped"—i.e., each cat has the top quarter-inch of his or her left ear removed. An eartipped cat trapped in error can easily be identified while still in the trap and immediately released. Also, if an eartipped cat is caught by animal control, it can be easily ID'd as a member of a managed colony and returned to the caretaker.

7. Once recovered from surgery, the cats are released as close as possible to where they were trapped.

My Own "Feral" Cats

The touchstones of the TNR method are spaying and neutering all cats in a feral colony and removing kittens (along with their moms) at an age early enough to socialize them to humans. The "window of socialization" in kittens comes and goes early and quickly: Between seven and thirteen weeks, the kitten learns whom (and what species) he can trust and live among.

Feral MomCats, generally too wild for socialization, are usually returned to their colonies, but only after they have weaned their kittens and been spayed.

Of my ten cats, all adopted from shelters, at least five started out life as ferals. My three oldest cats were rescued very early with their Mom-Cat, and they were fostered throughout kittenhood. They are extraordinarily people-friendly, well-socialized cats. My two youngest cats were born feral, trapped, and brought directly to a shelter at about ten weeks old. They were never fostered or handled by humans until that age. Now, at seven years old, Syvert the Bold is a determined snuggler, but he's still extremely jumpy and doesn't tolerate being picked up or carried.

His sister, Hazel-Marie, is a happy, healthy cat who gets along well with her cat family and even has a few snuggle buddies. But she hates being touched or handled by humans, which makes providing her with regular veterinary care rather challenging! Once in a while Hazel-Marie does let me stroke her, so I'm hoping that she will

My "feral" cats Hazel-Marie and Syvert the Bold.

be one of those rare feral cats who eventually warm up to both her human and feline family.

How to Get Started with TNR

1. Identify the problem. How many cats make up the colony? Learn all you possibly can about the cats, the site, and the community.

2. Educate yourself about TNR. Visit the websites of Alley Cat Allies, Alley Cat Rescue, Feral Friends, and others (see Resources). Call on these groups for advice and resources *before* you start planning.

3. Identify resources. Call your local shelters and get information about any active TNR volunteers in your area. Contact these people. Describe your colony and what you want to accomplish. Can you borrow equipment (such as traps) from them? Start locating the volunteers you'll need to trap, transport, and monitor the colony. Is there a low-cost or free spay/neuter service in the area, so the cats will not have to be transported a long distance? Find and make these contacts as soon as possible.

4. Involve the local community and communicate your intentions to them. Tell local residents and businesses about your plan to trap, neuter, and manage the colony cats. Explain how it will help both the cats and the community. You may have to overcome resistance from elements of the community, so do your research ahead of time and be ready for any questions or objections that might arise. Your goal is to recruit community members as partners and supporters. Expect this education process to take some time.

5. Plan! You'll need funds to cover food, spay/neuter costs, and any foul-weather shelters you plan to provide. The actual trapping and transportation will require many hands. Make a schedule and chart out all the different tasks that must be done. Who will be responsible for what? Who will visit the colony each day with food and water? Who will build and set up shelters for the cats? Get solid commitments from your volunteers and from your spay/neuter provider well in advance of trapping day.

6. Obtain humane traps. You often can borrow these from established TNR groups.

7. Set up feeding and water stations where the cats gather. Protect food and water from rain and weather. Devise a regular feeding schedule and get solid volunteer commitments for food visits. Make sure the colony

cats can expect at least one feeding visit per day, preferably at regular times.

8. Make sure the colony cats have accessible shelter, especially if they're subjected to rain, snow, extreme heat, or other severe weather. For example, you could use old dog houses with straw on the floors. Some groups build custom shelters for their colony cats; there are plans and instructions online (see Resources).

9. Schedule your spay/neuter appointments. Plan every detail carefully.

10. Trap each cat. This is harder than it sounds. Be patient, and always keep the welfare and safety of the cats uppermost in your mind. Trapping feral cats is more art than science, and experience helps. Tap into the experience and wisdom of national feral cat organizations online (see Resources) as well as local TNR experts. Bait the traps with especially tasty canned cat food or other meat. It's essential to check your traps often to minimize stress on the felines. Stay calm, even if the cats don't (and they won't). Covering trapped cats with blankets or towels can help.

11. Pick up the traps and transport the cats to the spay/neuter clinic. There, they will be spayed or neutered, vaccinated against rabies and other feline illnesses, treated for parasites, and possibly given other necessary medical care. Each cat will be eartipped for future identification.

12. Return each cat as close as possible to the exact location where he was trapped.

13. Now for the long haul: Oversee, monitor, and manage the established colony.

- Obtain firm commitments from your volunteer colony monitors for regular visits, feeding, retrapping, etc.

- Plan and conduct an ongoing TNR effort to neuter, vaccinate, and eartip cats who visit, including newcomers to the colony, as soon as they appear.

- Trap any pregnant cats so the moms-to-be can be fostered and given extra nutrition during pregnancy and kittening. Ideally, their kittens will be fostered and socialized to humans from birth and thus be highly adoptable. Most feral moms are unable to adjust to life as pets with humans, so they are spayed and returned to their colony. A few (especially very young MomCats) may prove adaptable enough to be tamed and socialized. These cats will stay in their foster homes for further socialization and eventual adoption.

- Keep an eye out for any cats who are tame enough to be placed for adoption. Strays (lost pets) often join up with feral colonies because of the consistent food source and other available resources.

- Provide fresh water and food daily.

- Observe all cats for any medical problems. Make sure cats with medical needs are retrapped and given the care they need as quickly as possible.

- Ensure that each cat is revaccinated at appropriate intervals.

- Provide the cats with shelter against harsh weather as necessary.

- Maintain a friendly dialogue with the local community. Listen to their concerns. In a timely, humane, and nonconfrontational manner, resolve any problems that pop up between the colony cats and their neighbors—human or otherwise.

- Maintain an ongoing program to recruit additional and substitute caretakers to ensure that the colony cats will always be cared for.

- Track the number of cats in the colony over time. Keep good records of all trapping, vaccinations, veterinary care, incoming cat numbers, and other interventions. This will prove valuable in explaining and proving the benefits of TNR to the local community, and to other communities considering the TNR approach. Be a model for others.

What about Feeding Strays?

Have you been feeding a few stray cats, who've come to know and trust you? Do the right thing: claim ownership of these cats. If you can safely approach and handle the cats you're feeding, and they're reasonably friendly, it should be simple to pack them off to your veterinarian for a complete checkup, spay or neuter surgery, and vaccinations.

If you've been feeding a large group of cats, and many or most of them are truly feral and unhandleable, don't try to resolve the problem alone. Help is available. Seek advice from humane organizations and cat rescue groups. (See Resources.)

Zoonoses: Rabies and Other Cat-borne Illnesses

It's always wise, if not always fair, to assume that any unknown cat is afflicted with a disease it could pass on to humans. Cats can carry a number of such *zoonoses*.

Most zoonoses are relatively difficult to catch. Common sense, wise cat-handling etiquette, and good hygiene will keep you and your family safe. Youngsters, the elderly, and immunocompromised persons are most at risk from serious complications of zoonoses.

VIRUSES AND SPECIES-SPECIFIC DISEASES

Let's clear one thing up right now: Although cats can suffer from a virus called FIV, often called "cat AIDS," this is *not* the same virus that causes AIDS in humans. *You cannot get AIDS, or the HIV virus, from your cat, nor can he get it from you.* Humans *cannot* get FIP (feline infectious peritonitis) or FeLV (feline leukemia virus) from cats, or from anything else. These diseases affect cats only.

You can't catch a cold from a cat, either, and no cat can catch one from you. Colds are species-specific.

RABIES

Rabies, a virus that affects warm-blooded animals, is by far the most serious threat posed by stray and feral cats. Rabies is transmitted in the saliva from the bite of an infected animal. It's virtually always fatal if not detected and treated immediately. In the United States, twice as many cats as dogs are confirmed as infected with rabies in a typical year. If you are bitten or scratched by a stray or feral cat, or any cat with unknown rabies vaccination status, it's critical to capture the animal for observation and testing. Post-exposure treatment of suspected rabies cases is less painful and traumatic than in times past, but it's still expensive and inconvenient.

OPCs and NCs, as well as your own outdoor cats, can pick up rabies in tussles with raccoons, foxes, infected cats, and other wildlife. Virtually all areas of the United States require cats to be vaccinated against rabies. Don't skip this important protection.

PLAGUE

Cat fleas can carry the bubonic plague organism, *Yersinia pestis*. Fleas have long been the bane of both humans and cats; the bubonic plague that wiped out one-third of Europe's population in the fourteenth century was carried by rat fleas. In the United States, plague is mainly a problem in the Southwest and California. Microbiologists have identified plague in flea-infested squirrels, chipmunks, and other wild rodents in several counties surrounding Sacramento. California health officials issue regular warnings about plague-prone areas in the state.

Cats can become infected with plague from flea bites, or by eating infected small mammals. Most people who get plague get it from a flea bite. Plague can also be transmitted by a bite, scratch, cough, or sneeze from an infected person or animal, including a cat. In recent years, almost all human cases of pneumonic plague, the worst kind, have been linked to cats.

CAT SCRATCH DISEASE

Each year, about twenty-two thousand cases of CSD (cat scratch disease, also called cat scratch fever), leading to about two thousand hospitalizations, are reported in the United States. CSD is a self-limiting, generally mild and benign infectious disorder that usually clears up on its own in two to five months, even without antibiotic therapy. It's relatively uncommon and not easily acquired. One episode appears to confer lifelong immunity in children and young adults. It's not contagious between humans.

Like plague, CSD is carried by fleas. The common cat flea transmits *Bartonella henselae* (the bacterium that causes CSD) to cats. In the absence of fleas, an infected cat's ordinary daily interactions won't transmit infection to uninfected cats. Infected cats seem to suffer few or no ill effects from the bacterium. Cats with fleas pose the highest risk for transmitting CSD to humans. However, the cat flea may be able to transmit *B. henselae* directly to humans—no cat needed.

Eighty percent to 90 percent of CSD patients are under twenty-one (many are under twelve), and males are affected 50 percent more often than females. While 80 percent of patients report earlier contact with a cat, cat-free CSD has been reported after contact with squirrels, dogs, goats, crab claws, and even barbed wire.

In a very small percentage of cases, serious problems can occur. Children with CSD should be vigilantly watched for complications:

- Perinaud's oculoglandular syndrome (when the cat scratch is on or near an eye, and includes inflammation of the eyes)

- Encephalopathy (with an onset about six weeks after lymph node swelling, and including fever, coma, or convulsions, or even transient blindness)

- Osteomyelitis (bone marrow infection)

- Pneumonia and pulmonary disease

Though serious, these complications generally clear up without permanent effects to the patient.

Common sense, good hygiene, and responsible cat-handling etiquette will prevent most cat-borne CSD. The infection-carrying cat bite or scratch is usually on the neck, head, hands, or forearms. Children who clutch flea-infested cats and kittens to their faces and chests, or play with them with their hands and arms, are most at risk.

RINGWORM

Ringworm has nothing to do with worms. It's a fungal skin infection, also called dermatophytosis, usually caused in cats by the fungus *Microsporum canis*. Other fungi, *Trichophyton mentagrophytes* (from contact with rodents) or *Microsporum gypseum/fulveum* (from spores in soil), more rarely cause cat ringworm. Ringworm is highly contagious among cats. Symptoms include reddish, scaly patches of missing fur, especially around the face and ears; scaly patches on the paws; and an overall sparse, moth-eaten coat.

Unlike most other zoonoses, ringworm is easily transmitted to people, showing up as athlete's foot, jock itch, ringworm of the scalp (scaling and hair loss), nail fungus, or rashes and skin lesions. Both children and adults can easily get ringworm playing with unknown and stray cats. Ringworm is devilishly difficult to eradicate and has a nasty habit of popping back up after you think it's finally gone. Ringworm spores are extremely resistant to water, heat, chemicals, detergents, and medication and can live in an ideal environment for up to eighteen months.

Got ringworm? Try a nonprescription antifungal cream. If it doesn't clear up in a week, see a physician. Cat got ringworm? Take him to the veterinarian immediately. Eradicating ringworm in a cat is a more complicated process, not a do-it-yourself venture.

Don't Underestimate Ringworm

Several years ago I adopted three kittens—Winnie, Jessie, and Pemi—from a local shelter. I always isolate incoming cats and kittens for a few weeks anyway, but in this case I was extra careful, as there had been a few cases of ringworm at the shelter. Sure enough, on their third day home, the kittens broke out with ringworm.

I immediately instituted strict isolation protocols between the kitten room and the rest of my house. I didn't want my other six cats to acquire this fungal infection, which can persist for quite a while in the environment, on many different kinds of surfaces. The microscopic fungal spores that transmit ringworm are particularly adept at working their way into carpeting, upholstered furniture, curtains and drapes, and other soft surfaces, where they can be very hard to eradicate.

My veterinarian prescribed a topical antifungal medication for the kittens. Each evening after dark, I thoroughly inspected each kitten with a portable black light and applied the medication to all affected areas. In addition, the kittens visited the veterinarian once a week for a more complete testing and inspection regimen.

Fortunately, my kitten room (like the rest of my house) has hard floors, but I also took additional steps to prevent the spread of the infection:

1. I dedicated one of my portable vacuum cleaners to the kitten room for daily vacuuming. It stayed there until the ringworm episode was completely over, at which point I decontaminated it, inside and out, with bleach.

2. Because ringworm spores can travel through heating and air-conditioning ducts, I blocked off the heating duct in the kitten room. (Fortunately, the weather was warm.)

ROUNDWORMS

Cats pick up roundworms by eating infected prey. Kittens can get them in Mom-Cat's milk. The most common cat worm, *Toxocara cati,* is a three- to five-inch-long parasite that lives in the small intestine and absorbs nutrients from the intestinal tract. A kitten with roundworms may have diarrhea, a potbelly, dull, rough fur, and flaky skin. He may cough and vomit frequently. Untreated roundworms can cause intestinal blockage and even death.

The greatest danger to humans who accidentally ingest roundworm eggs or larvae is visceral larva migrans, especially a risk for young children. A child may pick up the worm eggs from handling the feces of an infected cat in a sandbox,

3. I removed all nonessential items from the room and decontaminated them with bleach.

4. Each day I gathered up all washable items in the room into a plastic trash bag, transported them directly to my washing machine, and laundered them in hot water and bleach.

5. Each time I left the kitten room (leaving my clothing and shoes inside), I headed straight for the shower to wash up with chlorhexidine solution (obtained from my veterinarian). Chlorhexidine kills the ringworm fungus and is easier on skin than bleach.

6. Every few days the kittens were carried outdoors (in their carrier) for a sunbath in a cage. (My veterinarian had suggested that direct sunshine might help clear up the ringworm.) I used the opportunity to mop the floors and lower walls in the kitten room with bleach solution.

Sounds like I thought of everything, right? But I made one big mistake: I didn't remove the carpeted climbing tree/scratching post from the room right away. The kittens loved it so much I didn't have the heart to take it away from them. But despite spraying it and mopping it down with bleach solution, the tree turned out to be a stubbornly persistent reservoir of fungal spores.

When the kittens' ringworm hadn't cleared up after several weeks, I finally removed the tree, taking it directly outdoors for decontamination with bleach solution. After that the ringworm cleared up rapidly, and soon my veterinarian gave the go-ahead for the kittens to meet the rest of their new feline family. I had learned an important lesson: Never underestimate ringworm!

park, backyard, or indoor litter box, and then putting his fingers in his mouth. The eggs mature in the child's intestinal tract. Eggs that don't mature into adult worms can form cysts. The adult worms or cysts can migrate to various locations in the body, often the liver or the eyes. Blindness or other serious complications can result.

Because of potential risk to humans, researchers at the US Centers for Disease Control and Prevention say it's wise to assume that most, if not all, kittens and stray cats have roundworms. Modern treatments are safe and effective, so all kittens should be routinely dewormed.

HOOKWORMS

Hookworms are a type of roundworm that lives in the feline digestive tract. People and cats usually pick up this parasite from close contact with wet sand (sandboxes, beaches, beneath buildings). Hookworms can't complete their life cycle in human hosts, but humans can get a creeping skin eruption called *cutaneous larva migrans*. This can cause intense allergic reactions, such as inflammation and intense itching, followed by eruptions and itchy red wheals as the larvae migrate around the skin surface. It clears up on its own (usually within weeks or months), but the symptoms usually require treatment.

TOXOPLASMOSIS

Despite popular fears, owned cats are hardly ever the culprits in toxoplasmosis, an infectious disease that can cause serious injury or death to human fetuses. (It can also cause serious illness in children and immunocompromised persons.) In the United States, about three thousand people are diagnosed each year with toxoplasmosis. Most of them get it from eating raw or undercooked meat (especially lamb and pork) or unwashed vegetables, drinking unpasteurized milk, or working in contaminated soil without gloves. A study in the *New England Journal of Medicine* found *no* correlation between cat ownership and infection with toxoplasmosis.

NCs, OPCs, and outdoor cats can become infected with *Toxoplasma gondii* (the parasitical protozoan that causes toxoplasmosis) by eating raw meat (infected prey animals) or by contact with the feces of an infected cat. These are good reasons to keep unknown cats out of your gardens and keep your own cats indoors. Infected cats shed the parasite's eggs, or *oocysts,* in their feces for a very short time, but the oocysts can survive in the soil (under ideal conditions) for many months.

Strictly indoor cats hardly ever have the infection. Any indoor cat who does usually has it, and sheds oocysts, for only a few weeks, almost always during kittenhood. The risk of a pregnant woman acquiring toxoplasmosis simply by cleaning the litter box, or living with a cat, is greatly overblown. Anyone who's lived with cats for a significant length of time has probably already acquired immunity to toxoplasmosis.

Pregnant women worried about toxoplasmosis should wear rubber gloves and a face mask while scooping litter boxes, or delegate that chore to another family member. Simply cleaning the litter box at least once a day will also prevent infection. It takes at least a day for shed organisms to become infective.

De-skunking a Cat

If your cat ventures outdoors, and skunks inhabit your region (they're very common), sooner or later he might come home smelling less than . . . pristine. Traditionally a tomato juice bath, coupled with patience, the passage of time, and perhaps a clothespin on your nose, has been prescribed as the antidote to this unhappy condition. Unfortunately, the traditional treatment often doesn't vanquish the skunky odor, a remarkably persistent stench caused by chemicals called *thiols*. Plus, you may end up with a pink-tinged cat, something neither you nor your cat will find appealing.

Here's a better, quicker, and more inexpensive fix, developed by chemist Paul Krembaum. The solution he prepared from ordinary household ingredients causes oxygen molecules to bond with the thiols, effectively neutralizing their foul-smelling odor.

Packaging difficulties (the oxygen released by mixing hydrogen peroxide and baking soda can't be bottled, because it's likely to explode) led Krembaum to decide against trying to patent his formula. Instead, he made it available free of charge, a boon for desperate pet owners:

- 1 quart 3 percent hydrogen peroxide (very inexpensive; available at grocery, discount, and drug stores)

- ¼ cup baking soda (probably in your cupboard already)

- 1 teaspoon liquid soap (right by your sink)

The soap breaks up oils in the skunk spray so the other ingredients can neutralize the thiols.

Bathe your stinky cat in this mixture. Be sure to protect his eyes. After the bath, rinse off the solution with plenty of tap water. There's a small chance that the hydrogen peroxide in this mixture could bleach your cat's fur.

Warning! *Do not store* this mixture. It might explode.

TICK-BORNE DISEASES

Ticks carry several diseases. The most serious are Lyme disease, cytauxzoonosis (a rare parasitical infection, often fatal to infected cats), and Rocky Mountain spotted fever.

Lyme disease is rarely fatal, but can cause serious illness in humans. Wandering cats (OPCs, NCs, and your own) can pick up ticks, putting at potential risk anyone who subsequently handles them. Cats don't seem to be bothered by Lyme disease; they're just transmitters of the infective ticks.

A cat can also get Rocky Mountain spotted fever from an infected tick. It won't make *him* sick, but he can potentially pass it on to *you*.

LESS COMMON CAT-BORNE ZOONOSES

Cats can harbor a number of troublesome microbes that can be transmitted to humans, causing a range of intestinal problems. These include *Campylobacter enteritis* and *C. jejuni, Salmonella typhimurium,* and *Giardia lamblia*. Sensible cat etiquette (including avoiding contact with unknown cats), hand washing, and everyday good hygiene will keep you and your family safe.

A NOTE ON TAPEWORMS

Tapeworm eggs are not directly infectious to humans. They can affect your cat, though, if he has contact with infected cats or with fleas.

Cat (and dog) fleas are the intermediate host for the tapeworm *Dipylidium caninum,* the tapeworm that usually plagues cats. Flea larvae eat tapeworm eggs. Then a cat eats the flea as he tries to get it off his skin. The tapeworm ends up in the cat's digestive tract, where it can develop into an adult. In as little as two weeks, the tapeworm can settle in and start producing egg sacs. These short white egg cases (about ½ to ¾ inch long) can move about independently. They're usually seen around the cat's rectum, or in his feces. These egg cases aren't the tapeworms. The worm that sheds the sacs remains attached to the cat's intestinal wall.

Chapter 14

Helping Cats in Your Community and Beyond

A hallmark of responsible cat owners is their desire and willingness to look beyond their own feline families, to seek out and offer whatever assistance they can provide to cats in their communities and beyond who are in need of help. There are so many ways you can help cats! From answering a neighbor's questions about cat behavior to scooping litter boxes at your local shelter; from writing a donation check to bringing daily meals to a colony of feral cats; from doing paperwork in the shelter office to blogging about your cat-rescue activities; from interviewing potential shelter adopters to socializing and fostering kittens in your home; from writing "cat biographies" and taking pictures of adoptable cats to post on the Internet to offering humane education and responsible cat care information for local schools—there's a job for everyone. Most of all, though, each cat owner should strive to be a role model for other cat owners by providing their own cats with excellent, high-quality care throughout their lives.

Be a Role Model: Spay and Neuter Your Own Cats

To shelter workers across the United States, the coming of spring means just one thing: Kitten season is on the way. Soon they'll be hip deep in kittens for whom there are simply not enough adoptive homes. Though we've made tremendous progress in the past decades, pet overpopulation is still a stubborn, tragic problem. Each cat spayed or neutered helps. Wouldn't it be nice to see spring kitten season greeted with joy instead of dread?

Responsible cat owners spay and neuter their cats. Unless you're an experienced breeder involved in a responsible, planned breeding program designed for the continuing benefits of pedigreed animals, there's no good reason *not* to sterilize your cat.

Cat owners hear (and make) plenty of excuses. A common one is that having one litter makes a cat tamer and more people-friendly. According to experienced breeders, though, a fussy or prickly cat remains just as fussy or prickly after mother-hood. Pregnancy does induce hormonal changes that can make the MomCat-to-be *temporarily* cuddlier, but it won't change her underlying temperament or personal-ity. Another old favorite is that an altered cat will get fat and lazy. Again, there's a grain of truth here, but just a grain: Altered animals *do* need fewer calories. If you keep piling on the chow in the same old big portions and don't ensure that your cats get plenty of daily active play and exercise, yes, they will get chubby—just like us.

Spayed and neutered cats enjoy longer, healthier lives. Spaying your female cat eliminates her risk of uterine and ovarian cancer, birth complications (includ-ing cesarean sections), pyometra (a serious uterine infection), and ovarian cysts. She won't be burdened by repeated heat cycles, which are stressful for both cat and owner. If she's spayed before her first heat, she'll enjoy a greatly reduced risk of breast cancer. A spayed female cat, free of the complications of mating and reproduction, is a calmer, happier companion.

Neutering your male cat before he reaches sexual maturity reduces his risk of prostate enlargement, prostate cancer, and testicular cancer. Neutering will also make him more affectionate, less aggressive, less likely to roam, less likely to get in fights or get lost, and much less likely to spray and urine-mark in your house. The bite wounds, abscesses, torn ears, scratched eyes, infections, and other battle dam-age to a roaming, un-neutered tomcat can rapidly rack up hefty veterinary bills.

Unfortunately, there's still a very poor compliance rate in spaying and neuter-ing of cats adopted from shelters and rescue groups. Because of financial or emo-tional considerations, or belief in those old myths, too many owners still resist the very idea of altering, or just put it off too long. Oops! Another batch of kittens for the overburdened shelter.

Because of this, many shelters now routinely spay or neuter all kittens before adoption, a practice endorsed by the American Veterinary Medical Association. Pediatric spaying and neutering, also called "early age" or "juvenile" sterilization, involves spaying or neutering kittens as young as eight weeks and at least two pounds in weight, instead of waiting until just before or after sexual maturity.

Both pediatric and traditional-age spay and neuter surgeries are very safe. Complications are exceedingly rare. Healthy animals generally tolerate anesthe-sia well, are closely monitored during and after surgery, and usually go home the same day.

So, be a good role model. If *your* cat isn't yet spayed or neutered, what are you waiting for? Schedule that surgery. Don't be part of the problem. Be part of the solution: one person and one cat at a time.

Be a Resource for Others

By reading *Outsmarting Cats,* you've learned a lot about cats, cat behavior, and caring for our feline friends. You know what cats want and need, and you know how to persuade them, in cat-friendly ways, to do what *you* want, while thinking it was *their* idea all along. Share your new knowledge with others. Tell your friends about your favorite cat toys. Share cleanup tips. Is a neighbor talking about declawing because his cat is shredding the furniture? Tell him why declawing is a bad idea, and recommend Sticky Paws and a great new scratching post.

Is your sister's cat driving her nuts by missing the litter box? Tell her about some of the things you learned in chapter 5. Help her diagnose her cat's problem and suggest some simple changes she could make to help solve it: Add a larger litter box? Relocate the box to a more peaceful and secure location? It's amazing how often just adding a litter box solves a litter box issue, and keeps a cat in his home. Your advice can save feline lives.

Make Your Donations Count—Give Locally

If you're like me, you get lots of appeals in the mail from major animal charities. Although these organizations do some worthy work, most of these groups don't actually care for any animals. They don't operate a single shelter. I don't want my donations to fund mass mailings, big salaries, fundraising programs, or endless address labels. I want to get cats spayed and neutered, kittens vaccinated, hungry cats fed and sheltered. I want to see abused, abandoned, and homeless cats rescued and adopted into safe, loving, permanent homes.

Most small local animal shelters and rescue groups can't afford fancy mailings, or sometimes any mailings at all. I've observed that the best cat shelters are run on a shoestring budget by small bands of creative, courageous, tireless, low- or unpaid full-time workers—and lots of volunteers. They run yard sales and bake sales, and scrounge for donations of money and essential supplies. Everyone pitches in with the hard, often messy jobs: trapping feral cats, picking up strays, setting up spay and neuter appointments, transporting fractious cats, gentling feral cats so they can be adopted, fostering and socializing kittens, grooming and bathing, scooping endless litter boxes, and filling and cleaning endless food and water bowls. That's where I want my dollars to go!

Ask your veterinarian and your cat-loving friends: Who is doing good work for cats right here in our community? Chances are there are several small, local shelters, rescue groups, and TNR groups that could use your help. Visit their websites. Call them. Find out about their visions and philosophies. (An increasing number of shelters practice the "no-kill" approach, where no adoptable cat is ever euthanized.) Visit in person and get to know the people—and the cats. Find out what they need most. Most shelters and rescues have a wish list on their websites, listing everything from paper towels to computer equipment. Sign up to volunteer. Whatever your talents and abilities, just a few hours of your time a week can make a huge difference.

Then visit and see for yourself the good your donations can do when they're used wisely and efficiently by dedicated local people. Instead of tossing yet another sheet of address labels, you can enjoy the purrs and snuggles of your beneficiaries.

There's one exception I make to my "give locally" rule: the Winn Feline Foundation (www.WinnFelineHealth.org), a nonprofit organization that supports scientific studies of cat diseases and health problems. Despite Americans' love for their pets, basic scientific research that advances our knowledge of companion animal health and behavior receives only a tiny amount of funding and attention. Cat research in particular gets short shrift when it comes to funding. Yet a relatively small investment in basic research can lead to important veterinary medical breakthroughs, new treatments and medications, cures for previously incurable diseases, and other benefits for uncounted cats. Much of this research can also potentially benefit humans, as cats suffer from many of the same major and chronic afflictions that affect us, such as cancer, diabetes, and heart problems.

The Winn Feline Foundation takes on tough research problems and makes every penny count. For example, they're studying feline hypertrophic cardiomyopathy, an often fatal heart disease that affects many cats. The foundation is also dedicated to solving the heartbreaking puzzle of FIP (feline infectious peritonitis), a fatal feline disease for which there is currently no vaccine or cure.

Fostering and Socializing Kittens

Most shelters, rescue groups, and TNR groups have a continuing and urgent need for volunteer foster families. Fostering can be challenging, but it's one of the most rewarding things you can do to help cats in your community. Robin A. F. Olson, founder and president of Kitten Associates, a Connecticut cat rescue group, has fostered hundreds of cats and kittens. "It may seem like trying to empty the ocean with a thimble, with companion animal overpopulation into the millions," she says, "but to each cat and kitten you foster, it means the world."

It's a good idea to get to know the people at several local shelters and rescues, so you can find a good fit. Many organizations reimburse foster parents for veterinary and other expenses incurred, and you'll always have access to their experienced advice and help. As a feline foster parent, you're never alone.

GETTING STARTED IN FOSTERING

In a typical fostering scenario, a MomCat with kittens needs a safe, suitable place to stay while her kittens are young and vulnerable. Moms and kittens have special needs for peace and quiet, privacy, extra nourishment, close monitoring, and gentle handling. These needs are hard to meet in a busy, noisy shelter environment. As the kittens grow, they'll need to be socialized to humans, as well as to other cats and other pets, so that they'll be as adoptable as possible. Foster homes that have a lot in common with the cats' future permanent adoptive homes are the perfect solution. Ready to get started?

ASK A LOT OF QUESTIONS

You'll want to ask lots of questions, especially if you've never fostered before. How many kittens are there? When were they born? Is MomCat a feral cat from a colony? A stray recently brought to the shelter? Is she people-friendly, tame, and easy to handle? (Feral momcats usually require extra precautions and more experienced foster parents.) How well does she get along with other cats?

Even though your foster cats will be living in a space away from your own cats (ideally, a separate room), you'll need to ensure that your own cats won't be put at risk of contracting a disease or picking up parasites from the foster cats. Has Mom-Cat been tested for feline leukemia virus (FeLV) and feline immunodeficiency virus (FIV)? If MomCat has not tested negative for both viruses, what special precautions and care will be needed? Does MomCat have any other special needs? Has she been checked and/or treated for ringworm, fleas, worms, and other parasites?

Be aware of any state or local laws and regulations that might affect your fostering activities. For example, if the cats have come to you from out of state (for example, through one of the many multistate rescue networks around the United States), they will probably have to be checked by a veterinarian within a short time of their arrival—and possibly several times thereafter. The shelter or rescue group you're working with should have all the details. Be sure to ask.

PREPARE YOUR SPACE

The ideal fostering space is a separate room that can be closed off to keep other cats, other pets, and small children out. A spare bedroom, laundry room, or bathroom works well. MomCats need plenty of peace and quiet and privacy. The space should be clean, dry, warm, and free of drafts. Do as much kitten-proofing as possible before your foster family arrives. Block off access to any location where a kitten could become trapped, such as under and behind appliances and large pieces of furniture. Make sure there is no space where MomCat or the kittens could hide or become inaccessible. Remove all toxic and dangerous materials, and any objects that could fall and injure a kitten.

Prepare a nesting box. This can be a large, high-sided cardboard box, a large dog crate, or a cage. The sides of the box need to be low enough (or lowerable) so that the kittens can climb in and out on their own when they're old enough to start exploring. If your fostering room is a bathroom, you can place the nesting box on one end of the bathtub and the litter box at the other end. The nest box needs to be large enough for MomCat to stretch out comfortably and nurse her kittens. An open top will make it more convenient for you to check on the family and to pick up the kittens for weighing and inspection. Line the bottom with several layers of clean, washable towels or blankets. As the materials become soiled, you can simply remove them and expose a fresh surface. Change the bedding every day, even if it doesn't seem dirty.

If possible, place a tall climbing tree/scratching post in your fostering room. This will give MomCat something she can climb to get an occasional respite from her kittens (especially important when she is trying to wean them). It will also teach the kittens to use a scratching post.

MomCat will need an easily accessible litter box close by. A low-sided box is best, as the kittens will be using it, too, as soon as they start eating on their own. If you know what MomCat's litter preferences are, try to use her favorite litter at first. Your goal is to make MomCat feel as comfortable and safe as possible. However, please note that some kittens will try eating the litter. If it's a clumping litter, this can have serious side effects, including internal blockages. To be on the safe side, when the kittens start using the box, switch to a clay, non-clumping litter, and be vigilant for signs that any of the kittens are snacking on it.

WELCOME MOMCAT AND HER KITTENS

Depending on whether MomCat is a feral cat or tame and socialized, she will take anywhere from a few hours to a few days to settle in and become reasonably comfortable. Set everything up, place the open cat carrier in the room right by the nesting box, and leave for a while. It's a good idea to partially cover the nesting box with a blanket or towel, to make it seem extra cozy and appealing. Once MomCat feels safe, she'll move the kittens into the box. If she doesn't, gently transfer the kittens to the box. If MomCat's still nursing her kittens, make sure she resumes nursing as soon as possible after arriving. Make sure MomCat has well-filled, easily accessible food and water dishes available at all times.

If the kittens are being weaned or have been weaned, place a low-sided bowl of kitten milk-replacement formula, and a saucer of high-quality canned kitten food in the box. Watch each kitten to make sure he can eat and drink without choking. Some kittens may need a little help with this; wipe some of the food or milk onto their mouths or lips to give them the idea that "Hey, this is food!"

Keep a discreet eye on the family, but don't hover, especially at first. Move slowly, and talk softly and reassuringly to accustom the family to associating human voices with food and loving care. You can also leave a radio on low volume to further accustom the cats to human voices. Visit frequently to check on everyone, clean the litter box, and offer some gentle talk and handling.

For MomCat's safety and security, keep a night light on at night and, if there are no windows, a room light on during the day. Always keep the litter box spotlessly clean.

If the kittens are very young, MomCat may try to move them. This is a very strong instinct that, in the wild, helps protect the kittens from predators. The safer and more comfortable MomCat feels in the nesting box, the less inclined she'll be to move. Be a good provider and make the nesting box the most attractive place in the room, though, and MomCat will likely stay put. It's vital that the nesting area stay warm and dry, as kittens can get chilled quickly. If MomCat moves the kittens to an inappropriate spot, put them gently back in the warm nesting box.

TAKE CARE OF MOMCAT SO SHE CAN TAKE CARE OF HER BABIES

If the kittens are four weeks old or under, MomCat will be doing most of the work. Your job will be to take the best possible care of her, so she can take the best possible care of her babies. Nursing momcats need a *lot* of nourishment. Feed MomCat as much as she'll eat of a high-quality canned food. Provide plenty of fresh water, and leave some dry kibble in a bowl near her water dish. If MomCat seems reluctant to leave her kittens to eat, serve her food on a small saucer and place it near her head, so she can eat without getting up.

It's vital to ensure that MomCat is eating well, and that each of the kittens is actually nursing or eating. (It's sometimes hard to tell in the kitten pile-up.) For their first three weeks of life, the babies will nurse at least every one to three hours. After they're done, MomCat will lick their rear ends to stimulate excretion and then eat the output. This nest-cleaning activity is another defense against potential predators.

If Mom carries a particular kitten away from the group and seems to abandon it, she may feel that it is ill, or that it is going to die soon. This is a natural instinct. After seeking out the advice of your veterinarian, as well as your experienced fostering advisors, you may decide to try to bottle-feed the rejected kitten. This is a challenging, around-the-clock project that may or may not be successful.

IT'S NEVER TOO EARLY TO START SOCIALIZING

As soon as Mom seems settled in and calm, make it a priority to pick up each kitten for inspection and an initial weight check. Mom may fuss and seem very nervous when you handle her babies, but don't worry: She won't reject them just because you've handled them. Weigh the kittens at least daily, and record the weights. Each kitten should continue to gain weight steadily. If a kitten is not gaining, ask your veterinarian and fostering partners about possible bottle feeding and supplementation. If your kitten inspections reveal anything that seems "off," consult your veterinarian.

Can't tell the kittens apart? That's a common problem! Some foster parents put different-colored collars on the babies. (Make sure you check the collars' fit frequently, so they don't become too tight on the rapidly growing kittens.) Or take a few minutes to examine each kitten closely and look for any distinguishing characteristics—one little mark, the shape of the eyes, or even their personalities. As you gain experience, it will be easier to distinguish each individual kitten.

Start handling the kittens as soon as possible. The more loving, gentle handling the kittens receive as youngsters, the more tame, people-friendly, well-socialized, and adoptable they'll be. Many foster parents think that clipping the kittens' claws helps these early encounters go more smoothly. Ask your veterinarian or a veterinary technician for advice on clipping tiny kitten claws.

Lift each kitten gently (you can wrap him in a small soft towel first), and place him on a soft blanket in your lap. Stroke him gently while talking in a soothing voice. Try cuddling the kitten next to your bare skin, which kittens seem to find comforting. Gradually, you'll feel the kitten relax. Using a small, soft baby brush, brush the kitten's fur gently. This will help accustom him to regular grooming, and it will help him bond to his human caregivers. Make each encounter short but sweet, and make sure each kitten experiences your frequent, loving touch. Be sure to wash your hands before and after handling the kittens.

BE ALERT FOR PARASITES

If you see any signs of fleas on MomCat or the kittens, treat MomCat with a topical "spot on" flea treatment. Ask your veterinarian to recommend a safe product. Do not treat the kittens until your veterinarian has given you a clear go-ahead. Even cat-safe flea products can be dangerous for very young kittens. Fleas are not just an annoyance; young kittens with severe flea infestations can rapidly develop life-threatening anemia. You can use a small, narrow-tooth flea comb to remove fleas from the kittens.

Consult your veterinarian about timing and safe products for treating the kittens for worms (hookworms, roundworms, and tapeworms) and other internal and external parasites.

INTRODUCE THE KITTENS TO SOLID FOODS

When the kittens are about four weeks of age, Mom will start weaning them, and it'll be your turn to be their primary food provider. Start by setting out one or more low-sided bowls of kitten milk replacer in the nest box. Make sure each kitten starts to eat. This learning process can be messy, but keep the nest box and surrounding area as clean as possible. At five to six weeks, start adding platters of canned kitten food. Continue daily weight checks to ensure that each kitten is steadily gaining weight.

SAYING GOOD-BYE TO MOMCAT

If MomCat is a feral cat from an established colony, she'll need to be spayed and returned to her colony site as soon as possible; the longer she is away, the harder her readjustment will be. Once the kittens are fully weaned (usually at about six weeks), MomCat can be safely removed. It's beneficial to the kittens to have her

Using Social Media to Help Cats: Tips from the Experts

I asked several top social media gurus, pet experts, and cat bloggers how social media is helping cats right now, and how cat owners can use social media tools, such as blogging, Facebook, Twitter, Pinterest, and others, to help cats.

How do you use the Internet and social media to help cats?

- **Educate other pet owners** about proper cat care, cat health, and cat behavior. Social media greatly expands the reach of knowledgeable cat owners, cat experts, researchers, cat behaviorists, and veterinarians, helping them reach new and much larger audiences than would be possible through traditional media, such as print, TV, or radio. And cat owners now have opportunities to connect with experts they never would have been able to connect with otherwise.

- **Find homes for cats:** With photos and videos of adoptable cats available 24/7, adoptable cats have a much better chance of finding their forever homes. Sites such as PetFinder.com make it easy for a prospective pet owner to search for the perfect cat within her own community without even getting off her couch.

- **Get lost cats home again:** When a cat is lost, social media can get the word out quickly and efficiently, and can get large numbers of people looking for the cat in a very short period of time. *Tip:* When the cat has been found, be sure to update your status and spread the happy news.

- **Raise money:** Social media lets cat shelters, rescue groups, and other cat charities collect funds directly from their websites (with Wordpress widgets like ChipIn or through PayPal donations).

- **Help cats in immediate need:** When a cat needs help *right now,* having networks of concerned cat lovers already in place can mean the difference between life and death.

What are some DO's and DON'Ts for using social media to help cats?

- DON'T let the Internet replace regular veterinary care for your cat.

- DO strive to be a *reliable* source of information.

- DO seek out *credible* sources of information, such as sites ending in ".edu" and sources recommended by your own veterinarian. There is a lot of bad information out there!

- DON'T pass along opinions or rumors as fact. Check everything out first.

- DON'T believe everything you read, and don't share stories or re-tweet without checking out the information through a trusted source.

- DO join Facebook fan pages of credible sources of information: veterinary colleges, American Veterinary Society of Animal Behavior (AVSAB), American Veterinary Medical Association (AVMA), American Association of Feline Practitioners (AAFP), and trusted pet journalists.

- DO work all the time to build your own networks and extend your reach.

- DO make friends with other cat owners, cat bloggers, and cat rescue folks. The more people you know, the more people you can reach out to when you have a specific cat in need of help or a cat who needs help fast.

- DO join networks such as BlogPaws (www.BlogPaws.com) to partner with other cat bloggers to help spread the word about cats in need.

- DO cross-post your *reliable* cat news, information, video, or photos anywhere and everywhere. Use Facebook and Twitter, of course, but don't forget about Google+, LinkedIn, Blogger, Tumblr, Stumbleupon, Digg, Reddit, Pinterest, and all the others.

- DO ask others to re-post, re-tweet, or blog about your *reliable* cat story, article, video, or photo. Cat-loving friends are always willing to help spread the word when it comes to helping our feline friends.

- DO accentuate the positive.

- DO keep your posts and tweets factual, upbeat, and positive.

- DON'T share graphic photos of animal abuse. Many people just can't handle seeing those awful photos in their news feed day after day.

- DON'T pass along gloomy, pitiful stories about cats on death row, cats with severe physical challenges, etc. It's just not an effective way to find homes for animals. Studies show that if animals are presented in a positive manner, no matter what their challenges are, they're much more likely to find permanent homes.

- DO be innovative and creative.

- DO use YouTube videos and similar tools to showcase cats up for adoption.

- DON'T refuse Twitter friends who tweet as their cat. They're among the most devoted cat owners!

- DO make personal comments when appropriate in your tweets.

- DO send out useful tweets with good information in short bites.

- DO tweet something fun every so often, such as a quote from someone famous that pertains to cats.

- DON'T forget that social media is *social.*

- DO remember to be gracious. It's not about you; it's about the cats!

- DON'T use every tweet or post to push your own cause or sell something. A good rule of thumb: Write two posts about others' causes or content for every one post about your own.

- DO comment on other bloggers' posts, photos, and videos. Social media is, above all else, a *conversation.* Make it a conversation about cats.

The Experts

- **Darlene Arden,** pet expert, author, and blogger; www.darlenearden.com

- **Angie Bailey,** award-winning cat blogger; http://catladyland.net

- **Steve Dale,** award-winning pet journalist and blogger, www.stevedalepetworld.com

- **Jenny Dean,** Ragdoll cat expert and cat blogger; www.floppycats.com

- **Lorie Huston, DVM,** expert on social media and blogging for pets; www.social-savvy-pets.com

- **Ingrid King,** award-winning pet author and blogger; www.consciouscat.net

- **Robin A. F. Olson,** award-winning cat blogger; www.CoveredInCatHair.com

- **Amy Shojai,** pet expert, author, and blogger; www.amyshojai.com

stay longer, of course, but you need to look at the trade-offs between MomCat's welfare and the kittens' health. A few feral momcats turn out to be surprisingly adaptable, showing clear signs that they may have a future as domesticated pets. Encourage this, but don't be surprised if it doesn't happen.

VISITING THE VETERINARIAN

According to the AAFP's vaccine guidelines, kittens can get their first vaccinations as early as six weeks of age. The initial vaccinations protect them against panleukopenia virus (FPV), feline herpesvirus-1, and feline calicivirus (FHV-1/FCV). They should be revaccinated every three to four weeks until they reach the age of sixteen weeks. Kittens can get their first rabies vaccination at either eight or twelve weeks, depending on the type of vaccine used. A kitten's first veterinarian visits should be as stress-free and pleasant as possible, offering many opportunities to accustom the kitten to handling by strangers, traveling in a carrier, and riding in a car.

SOCIALIZING THE KITTENS

Socializing your foster kittens begins, ideally, the first hour they're in your home. Plenty of gentle handling and soft, reassuring conversation will help prepare the kittens for their lives as domestic companions.

At about six to seven weeks of age, the kittens enter a seven-week period of maximum emotional and psychological flexibility. During this short "magic window of socialization," the kittens are extraordinarily open to meeting new people and enjoying new experiences. Their adaptability and learning abilities are at their height. As they are being weaned, they'll start showing a growing interest in their environment. They'll start wandering away from MomCat and trying to climb out of the nesting box. They are ready to explore their world!

Make sure they can get out of the box (and back in) easily and safely. Take time to review your kitten-proofing. Then, bring on the toys! Playtime is an excellent way to bond with the kittens and help them associate fun with humans. Provide small toys they can bat around on their own. Use interactive toys such as the Feline Fisher (see Resources) to lure them out of the box for some chasing and leaping. Conduct interactive play sessions several times a day, and intersperse these with plenty of snuggling, grooming, and lap time.

This is also the best time to invite friends over to cuddle and play with the kittens. The more people the kittens meet during this time, the more adaptable they will become to meeting new people in the future. Inviting children over to play with the kittens will help make them more adoptable; just be sure to monitor kid–kitten interactions. Use the opportunity to teach children safe and proper cat-handling techniques.

As the kittens become bolder, they'll venture farther from the nesting box, and for longer periods of time. Make sure, though, that they can always get back easily into the box. It's essential that they feel they always have a safe, familiar place to retreat.

All this exploration will make the kittens hungry. Be a good provider and be ready for them. Milk replacer and canned kitten food should be available at all times. Let the kittens eat as much as they like, and continue to monitor their weights.

MEETING YOUR FELINE FAMILY

Some foster parents keep their foster cats completely separate from their own cats. There are many good reasons for this, including preventing transmission of diseases and parasites, and minimizing disruption to the family cats. This is especially true if you have cats who are particularly sensitive to stress and changes. Being forced to meet and interact with several unfamiliar felines can send some sensitive kitties around the bend. In any case, it's vital to keep the kittens completely isolated from your own cats until they're at least sixteen weeks (four months) old, fully vaccinated, and disease- and parasite-free.

If your cats are willing, though, letting the kittens mix and mingle with them can have several important benefits. The kittens will learn to live with and interact with cats other than their mom and siblings, and your cats can help civilize them, teaching them proper feline etiquette and manners. The kittens can learn such important lessons as using the scratching post, staying off kitchen counters, and dealing with the dreaded vacuum cleaner monster—all skills that will make them more adaptable and adoptable.

SAYING GOOD-BYE

Most shelters, rescues, and other groups that use foster homes allow the foster parents to have as much or as little involvement in the adoption process as they want. You may want to help interview prospective adopters, or you may want to leave this task to shelter personnel. Perhaps you can help post pictures and descriptions of the kittens to such sites as PetFinder.com, where they will be seen by thousands of potential adopters. Many kittens are returned to their shelters to be placed for adoption, while others are adopted directly out of their foster homes. Whatever the procedure, though, at some point you will have to say good-bye to the kittens you've so carefully nurtured.

Foster parents agree: Saying good-bye is the hardest part of fostering. But even with tears in your eyes, you can be proud that you played such a vital part in nurturing these precious lives and readying them for adoption into permanent, loving homes. And with your fostering room now empty, you're ready to host a new feline family!

Reaching Out to the Global Cat Community

The Internet may just be the best thing that ever happened to cats. Not since their heyday in ancient Egypt have cats been so popular, visible, and important. Right now millions of people around the world are watching cat videos on YouTube, "tweeting" as their cats, blogging about their cats, posting cat pictures to Flickr, and updating their cats' status on Facebook or Catster.

It's easy to join the global cat community! Tweet your cat news on Twitter. Start a Facebook page for your cat. Share your cat knowledge and experiences in your own cat blog. It's easy to get started, and with services like Wordpress (www.wordpress.com) and Blogspot (www.blogger.com), you can be set up and blogging within minutes.

It's not all just for fun, though—thousands of cats owe their lives to the global cat community that meets every day on the Internet and social media sites. When a cat needs rescuing, word gets out on social media, and potential rescuers and adopters step up. Transportation is arranged, and the cat is delivered to his new home. Multiply this by the thousands of cats whose lives have been touched and saved by dedicated, generous cat lovers all over the world. And you can be a part of it! (See box on page 334.)

Chapter 15

Better Safe Than Sorry

Is This an Emergency?

Cats are notorious for hiding pain, injuries, and illness, even when they're severely hurt or seriously ill. It's a hardwired survival trait. A cat in the wild who shows that she's weak or ill is likely to have her territory, and the resources necessary for her survival, seized by a rival. Part of outsmarting your cat is keeping her healthy and alive—even when she doesn't cooperate.

The following symptoms *virtually always* indicate a medical emergency. If your cat has any of these signs, call your veterinarian, after-hours pager, or the emergency veterinary clinic immediately.

- Unconsciousness

- Any major trauma, even if the cat seems fine (hit by car, a fight with another cat or other animal, a serious fall, any accident)

- Breathing difficulty, including cessation of breathing, a blue tongue, extreme shortness of breath, noisy breathing, or sustained panting

If you're concerned about your cat, always call your veterinarian. Err on the side of caution. Don't worry about sounding like a nervous Nellie. Get over it. Do what's right for your cat.

- Bleeding from any part of the body, especially if it doesn't stop by itself within a few minutes

- Difficulty urinating, especially if accompanied by vomiting or lethargy (could be a potentially fatal blockage)

- Open wound or gash that is large or bleeds heavily

- Evidence of electric shock, such as electrical burns in the cat's mouth or on her paws, especially in the presence of chewed electrical cords

- Any injury or hit to the head

- Broken bones, either obviously broken or signs such as lameness, dragging of a limb, or difficulty bearing weight or walking

- Sudden onset of weakness or extreme listlessness, especially if accompanied by dilated pupils, rapid heartbeat, or shallow breathing

- Sudden paralysis of one or both rear legs. This can be evidence of a *saddle thrombus*, or feline aortic thromboembolism. *This is an emergency of the highest order, and requires extremely rapid veterinary intervention if the cat is to survive.* In cases of saddle thrombus, the cat's heart throws off a blood clot that lodges at the point in the aorta where it divides to supply blood to the rear limbs. Even with rapid emergency care, many affected cats die or never regain the use of their hind legs.

- Seizures, staggering, circling, collapsing, uncoordinated movements, loss of balance, mental confusion

- Dry, sticky gums (possible dehydration)

- Loose skin that doesn't snap back (possible dehydration)

- Heatstroke or heat exhaustion

- Poisoning; any evidence that the cat has ingested a poison, whether a plant, chemical, or other substance

- Straining and crying in litter box, whether depositing nothing or tiny bits of urine

- Sudden-onset squinting or holding an eye tightly shut; sudden copious flow of tears or any drainage from the eye

- Pupils conspicuously different sizes

- Evidence of pain, including crouching, crying, reluctance to be touched, or unusual aggressiveness or other reaction when touched

- Persistent diarrhea that lasts for more than a few hours

- Bloody diarrhea

- Bloody urine

- Persistent or copious vomiting, or vomiting accompanied by diarrhea, unusual behavior, or other changes

- Extreme or sudden behavior or mood change, such as violent attacks or aggressiveness

- Visible third eyelid. In an awake cat, no more than the tiniest sliver of this membrane should be visible.

Warning: Lovely Lilies Are Cat-killers

Beware of bringing home that pot of lovely Easter lilies. They can kill your cat.

Easter lily (*Lilium longiflorum*), tiger lily (*L. tigrinum*), rubrum (*L. speciosum*), Japanese show lily (*L. lancifolium*), various daylilies (*Hemerocallis* species), and the popular Stargazer lilies can cause rapid, irreversible kidney failure and death in cats. Nobody yet knows which specific toxin in lilies causes the destruction of the cat's renal tubular epithelial cells (cells lining part of the kidney). There's no known antidote.

All parts of these lilies are highly toxic to cats. Clinical experience shows that 50 percent to 100 percent of cats poisoned by lilies die from the poisoning.

It takes only a few nibbles of either flowers or leaves. Within a few hours, your cat may vomit, become lethargic, or lose her appetite. These symptoms continue and worsen as the kidney damage progresses. Unless treated aggressively within eighteen hours, the damage is irreversible. Your cat's kidneys will fail, and she'll die within thirty-six to seventy-two hours.

Prompt, aggressive veterinary treatment is essential. If you know or suspect your cat has consumed any part of a lily plant (or any other plant you don't know to be safe), rush her to the veterinarian immediately, especially if she's vomiting or lethargic. If you can, take the plant with you so your veterinarian can identify it more easily and initiate the most effective treatment. If aggressive therapy is begun in time, the affected kidney cells may regenerate and your cat may recover completely.

Sudden Confusion Might Mean Sudden Blindness

Your cat is suddenly bumping into things. She howls and cries, stumbling and falling in confusion. Her pupils are widely dilated and she looks terrified. She's trying to hide but doesn't seem to know where she is. What on earth is wrong? She may have gone suddenly blind—quite often the first sign of hypertension, or high blood pressure. The increased pressure in the blood vessels of her eyes has probably caused a retinal detachment. This is an emergency!

For your cat to have a chance to see again, she needs to be hospitalized and treated aggressively with antihypertensive drugs. If her blood pressure isn't brought under control as rapidly as possible (within hours), there's much less hope that the retinal detachment will repair itself. With immediate, aggressive treatment, most affected cats will reattach their retinas, probably within a few days. But the longer the retina remains detached, the less likely the cat will regain vision.

Your poisoned cat will be given general supportive therapy, including administration of fluids. This treatment is very effective if rendered within six hours of ingestion of the toxic plants. But your cat's odds of recovery go way down if treatment starts more than eighteen hours after ingestion.

Other lily plants also pose serious dangers to felines. Plants of the *Spathiphyllum* genus and calla lilies contain toxic oxalates. Lily of the valley (*Convallaria majalis*) harbors a cardiac toxin.

No other species (including dogs or laboratory animals, such as mice and rats) is known to be affected by the toxins in lilies. If there's a lily plant or flower in your home (or in your yard or garden, if your cat ventures outdoors), there's always the chance that she could be poisoned. If you have a cat, ban all plants and flowers of the Liliaceae family from your home. If you live with cats and enjoy decorating with plants and flowers, use safer alternatives, such as orchids, daisies, or violets.

Quick Tips

1. Cats and candles don't mix. The US Consumer Product Safety Commission reports that deaths from residential fires fell from 4,500 in 1980 to 2,660 in 1998, while candle-related fire deaths increased from

Things That Are Pretty Much Incompatible with Safe, Harmonious Life with a Cat

If you have a cat, and you have any of these in your home, you're asking for trouble:

- Lit candles or any open flame

- Small pocket pets (gerbils, mice, rats, chinchillas)

- Many varieties of common houseplants and flowers (see the list below)

- Dried flower arrangements

- Breakable items of sentimental or pecuniary value displayed anywhere except behind secure doors

Things you probably have to have in your home but that require constant vigilance:

- Mechanical furniture, such as recliners and sofa beds (these can kill cats)

- Electrical cords (monitor for any tendency to chew)

- Stovetops and ovens

- Cleaning products and other chemicals (stow in cabinets with locks or child-proof latches)

- Cords for draperies, venetian blinds, or mini blinds (cut or tie up safely)

- Plastic bags with handles (can cause trapping and suffocation—cut handles)

- Cat-toxic medications, such as Tylenol and plain aspirin (stow in cabinets with locks or child-proof latches)

20 to 170. Thirty-eight percent of candle fires occurred after candles were left unattended, abandoned, or inadequately controlled.

2. If you crave candlelight, use safe, electric (rechargeable or battery-powered) candles. (See Resources.)

3. Install smoke detectors on each level in your home, and outside each sleeping area. Test them monthly, and change batteries every six months (when you reset clocks in spring and fall).

Toxic, Irritating, and Dangerous Plants for Cats

Don't consider this list comprehensive. If in doubt about any unknown plant, consider it potentially dangerous to your cat. Ask your veterinarian, or a local florist or greenhouse, whether a plant you have any doubt about is poisonous to cats.

Alfalfa
Almond (pits)
Amaryllis
Apple (seeds)
Apricot (pits)
Arrow grass
Autumn crocus
Avocado
Azalea
Baneberry
Bayonet
Bear grass
Beech
Belladonna
Bird-of-paradise
Bittersweet
Black-eyed
  Susan
Black locust
Bleeding heart
Bloodroot
Bluebonnet
Box
Boxwood
Buckeyes
Burning bush
Buttercup
Cactus
Caladium
Castor bean
Cherry (pits)
Chinaberry
Chives
Christmas rose
Chrysanthemum
Clematis
Cornflower

Crown-of-thorns
Cyclamen
Daffodil
Datura
Deadly night-
  shade
Delphinium
Dicentra
Dieffenbachia
Dumbcane
Easter lily
Eggplant
Elderberry
Elephant's ear
English ivy
Euonymus
Evergreen
Fern
Flax
Four-o'clock
Foxglove
Golden-chain
Golden glow
Gopher purge
Ground cherry
Hellebore
Hemlock
Henbane
Holly
Honeysuckle
Horse chestnut
Hyacinth
Hydrangea
Indian tobacco
Iris
Jack-in-the-
  pulpit

Java beans
Jerusalem
  cherry
Jessamine
Jimsonweed
Jonquil
Lantana
Larkspur
Laurel
Lily (especially
  Easter lily,
  but also tiger
  lily, daylily,
  Stargazer and
  similar lilies,
  Asiatic lily, and
  rubrum)
Lily of the valley
Locoweed
Lupine
Marigold
Marijuana
Mescal bean
Mistletoe
Mock orange
Monkshood
Morning glory
Mountain laurel
Mushroom
Narcissus
Nightshade
Oleander
Onions
Peach (pits)
Peony
Periwinkle
Philodendron

Pimpernel
Poinciana
Poinsettia
Poison ivy
Poison oak
Pokeweed
Poppy
Potato (stems
  and leaves;
  skin, especially
  if greenish)
Privet
Rhododendron
Rhubarb
Rosary pea
Rubber plant
Scotch broom
Skunk cabbage
Snowdrops
Snow-on-the-
  mountain
Staggerbush
Star-of-
  Bethlehem
Sweet pea
Tansy
Tobacco
Tomato (foliage)
Tulip
Tung tree
Virginia creeper
Water hemlock
Weeping fig
Wisteria
Yews

Be Prepared for Emergencies

Keep a cat-safety kit handy for emergencies. It should include:

- A sturdy cat carrier (for each of your cats)

- Several large, soft towels

- Rubber gloves

- Forceps (to remove stingers)

- A fresh bottle of hydrogen peroxide 3 percent (USP)

- One or more cans of soft cat food

- Turkey baster, bulb syringe, or large medicine syringe

- Saline eye solution (to flush out eyes)

- Artificial tear gel to lubricate eyes after flushing

- Mild grease-cutting dishwashing liquid (for bathing your cat if her skin gets contaminated with a toxic substance)

The ASPCA's Animal Poison Control Center

If you suspect your cat has gotten into a dangerous plant, or any other toxic substance, or if she's showing any signs consistent with poisoning:

- Copious vomiting

- Lethargy

- Dizziness, confusion, walking in circles

- Foaming at the mouth

- Seizures

4. Never leave a cat unattended around any space heater.

5. Avoid halogen lamps. They become extremely hot, and tend to be very lightweight—easy for a leaping cat to overturn. If you must have them, keep them away from flammable drapes, bedclothes, papers, books, and low ceilings. Never leave them on when you're away from home.

6. Woodstoves, favorites of warmth-loving cats, cause many chimney fires. If you use your woodstove regularly, have the chimney cleaned at least twice a year.

- Losing consciousness

- Unconsciousness

- Difficulty breathing

—don't waste a second. First, call your veterinarian. Then grab your credit card and call the ASPCA's Animal Poison Control Center (APCC) immediately. (Don't call human poison control. They don't have the specialized information necessary to give reliable, instant advice in veterinary cases. Many medications and other substances safe for humans are very dangerous for cats.)

The APCC hotline is available 24/7. Trained veterinary toxicologists have the latest scientific journals and books, databases of poisons and antidotes, and extensive experience from thousands of animal poisoning cases.

Write this number by every phone in your home:

(888) 4ANI-HELP (426-4435)

You can pay for APCC's services with Visa, MasterCard, Discover Card, or American Express. They'll do as many follow-up calls as necessary in critical cases. If you ask, they'll contact your veterinarian. They'll also fax or e-mail, to you or to your veterinarian, specialized treatment instructions and current literature citations.

When you call, the specialist on duty will ask you for enough information to rapidly help you and your cat: your name, address, and phone number; the species, breed, age, sex, weight, and number of animals exposed to a toxic agent; the time since exposure; the exact nature, including brand name, of the toxic agent; and a detailed description of your cat's symptoms.

The ASPCA's APCC also maintains an informative website at www.aspca .org/Pet-care/poison-control.

You'll find general information about poisoning prevention, indoor and outdoor hazards, and a suggested home first-aid kit. You'll learn about common household toxins and hazards in your kitchen, bathroom, garage, garden, and yard. There's lots of information on poisonous plants, foods, medications, and household chemicals.

7. Many older homes, crawl spaces, and trailers have pipes wrapped with electric heating tape, which falls apart as it ages. If your cats (or NCs or OPCs) play, hide, or shelter beneath such structures, they might shred and further damage the insulation. (Mice, squirrels, and raccoons can also cause damage.) Once the insulation is gone and the wires are bare, they can overheat or short out, causing a fire.

8. Keep sturdy carriers, one for each cat, easily accessible, preferably near a door.

9. Place SAVE MY PETS stickers, indicating number and species of pets inside, on doors and windows, in case a fire or other disaster occurs when you're not there to direct rescuers to your animals. Stickers are available from the American Society for the Prevention of Cruelty to Animals (ASPCA), Humane Society of the United States (HSUS), animal shelters, and pet supply retailers. *Keep stickers updated so firefighters don't risk their lives trying to save animals who aren't there.*

10. If you call 911 to report a fire or disaster, tell the dispatcher if there are animals (and what kinds) trapped or at risk.

11. *Never* go back into a burning structure to rescue pets or other animals.

12. No matter how well you know your cat, don't expect her to act normally or rationally during a fire, emergency, or disaster. Expect the unexpected, and be ready.

13. When evacuating your cats during a fire or other disaster, remember that, by instinct, cats faced with danger seek low, protected spaces like closets, under beds, or cellars.

14. Get to know local rescue workers and firefighters. Many cats (and other animals) have been saved because when the alarm came in, responding rescuers knew the homeowners and whether they had animals in their home.

15. If your cat has ever showed any inclination to chew cords or wires, unplug power to all nonessential appliances when you're away from home.

16. Run phone, computer, and electrical cords and cables through cable guides, cord winders, or other protective structures.

Resources

Cat Urine and Feces Cleanup

NON-ENZYME CLEANERS

Non-enzyme odor cleaners use a variety of odor eradication methods, such as oxidation, chemical bonding, odor encapsulation, and essential oils. Some products use a combination of approaches.

ZeroOdor
(This is my favorite. Easy and neat to use; uses chemical bonding to neutralize odor molecules.)
www.zeroodor.com
www.zeroodorpet.com
Facebook: Zero Odor
E-mail: support@zeroodor.com
(800) 526-2967
Available online and in some retail locations

Odor Medic
(Based on essential oils.)
www.odormedic.com
E-mail: sales@odormedic.com
Odormedic, LLC
Minneapolis, MN 55331
(612) 270-2953 (8 a.m. to 7 p.m. Central for free consultation)

OdorXit
(Uses chemical bonding and absorption; some products use oxidation. Can require repeated applications, plus up to weeks to be completely effective.)

www.odorxit.com
(877) ODOR-XIT (636-7948) (toll-free)

StinkFree
(Uses oxidation. Must foam to the surface for repeated blotting.)
www.stinkfree.com
E-mail: customerservice@stinkfree.com
StinkFree, Inc.
7803 N. Kickapoo St.
Shawnee, OK 74804
(800) 824-5363 (toll-free)

ENZYME CLEANERS

Enzyme cleaners generally contain both enzymes and live bacteria; therefore, they have a limited shelf life. They should never be exposed to extreme temperatures, which can kill the live organisms and render the product ineffective. Always check the shelf life, and use fresh product if possible.

Dr. Elsey's Urine Removal Program with Aromatherapy
(Bacterial/enzymatic action.)
www.preciouscat.com
Precious Cat, Inc.
4201 S. Navajo, Unit 1
Englewood, CO 80110
(877) 311-2287 (toll-free)

Anti Icky Poo
(Active live bacteria and enzymes. Easy to use. Shelf life promised over two years; company also sells accessories, such as black lights and carpet injectors.)
www.antiickypoo.com
www.mistermax.com
Facebook: Anti Icky Poo
(800) 745-1671 (toll-free)
Also sold by CatFaeries: www.catfaeries.com

Urine Off

(Bio-enzymatic technology.)
www.urineoff.com
E-mail: customerservice@urineoff.com
Bio-Pro Research, LLC
1140 Tate Blvd. SE
Hickory, NC 28602
(877) URINEOFF (877-874-6363) (toll-free; sales and customer service 8 a.m. to 5 p.m. Eastern)
Available at many retail locations

Pure Ayre

(Plant-derived enzymes and essential oils. Easy to use. Food-grade safe; can be used on multiple surfaces and in washing machine. Needs oxygen to work.)
www.pureayre.com
www.caturine.com
Email: info@pureayre.com; info@thepureayrestore.com
Clean Earth, Inc.
21809 95th Place S
Kent, WA 98031
(877) 787-3297 (toll-free)

SCOE10X

(Uses beneficial probiotics. Needs rinsing and extracting after use.)
www.scoe10x.com
BioFOG, Inc.
5950 Shiloh Rd. East, Suite H
Alpharetta, GA 30005
(877) 648-4015 (toll-free)

Nature's Miracle Just for Cats Urine Destroyer (and other Nature's Miracle products)

(Three-year shelf life. Uses enzymes; takes up to two weeks to work as it air dries.)
Spectrum Brands, United Pet Group
www.EightInOnePet.com
E-mail: support@ilovenaturesmiracle.com
Widely available at pet stores, home and garden centers, home care catalogs, and pet care catalogs

LAUNDRY PRODUCTS THAT REMOVE CAT URINE ODOR

Febreze Laundry Odor Eliminator
(My favorite—easy to use and really gets the odor completely out.)
www.newfebreze.com
Widely available at retail stores

CLEANING ACCESSORIES

Stink-Finder UV Light and Flashlight
(Ultraviolet or black lights; flashlights.)
www.stinkfree.com
email: customerservice@stinkfree.com
StinkFree, Inc.
7803 N. Kickapoo St.
Shawnee, OK 74804
(800) 824-5363
Available at PetSmart, PetCo, and other pet supply retailers, as well as pet supply catalogs

Stainless steel litter scoop
(The best!)
www.durascoop.com
Sweet Home Industries, LLC
PO Box 943
Batavia, IL 60510
(630) 879-3579
Widely available at pet retailers

Vibrating PowerScoop
(Battery-powered litter scoop.)
www.petcrew.net/Powerscoop/power-scoop.htm
E-mail: info@petcrew.net
Ethical Products, Inc.
Fashion Pet, Inc.
27 Federal Plaza
Bloomfield, NJ 07003
(800) 223-7768 (toll-free)
Also available at PetSmart

CLEANING APPLIANCES

DeLonghi Steam Cleaner
(Easy, safe cleaning for floors, upholstery, and more.)
DeLonghi America, Inc.
www.delonghiusa.com
E-mail: customerservice@delonghiusa.com
Fox International
23600 Aurora Rd.
Bedford Heights, OH 44146
(800) 322-3848 (toll-free)

Oreck vacuum cleaners
(The absolute best for picking up cat hair and scattered litter.)
www.oreck.com
Oreck Corporation
1400 Salem Rd.
Cookeville, TN 38506
(800) 989-3535 (toll-free customer service)
(800) 819-8549 (toll-free ordering 24/7)

THE BEST CAT LITTERS

Dr. Elsey's Precious Cat Ultra
(Our favorite—unscented, and great for multicat households.)
www.preciouscat.com
4201 S. Navajo, Unit 1
Englewood, CO 80110
(877) 311-2287 (toll-free)
Available at PetSmart, PetCo, and other pet supply retailers

Also:

Dr. Elsey's Cat Attract Litter
(Brings cats who have been "thinking outside the box" back to the litter box.)

Dr. Elsey's Respiratory Relief Litter
(For cats who are allergic to other litters—available in clumping clay or silica gel forms.)

Dr. Elsey's Senior Cat Litter
(Finer grained, more comfortable on paws.)
Dr. Elsey's also makes specialized litters specifically for longhaired cats, kittens, senior cats, and cats recovering from surgery.)
Dr. Elsey's products are widely available at pet retailers.

World's Best Cat Litter
(Made from ground whole-kernel corn.)
www.worldsbestcatlitter.com
1600 Oregon St.
Muscatine, IA 52761
(877) 367-9225 (toll-free)
Available at many retail locations

Cat Hair Cleanup

The LolaRola
(Washable "sticky roller" mop.)
www.lolaproducts.com
E-mail: info@lolaproducts.com
Lola Products
343 S. River St.
Hackensack, NJ 07601
(800) 524-2822 (toll-free)
Available at numerous retail outlets

Sticky Critter (for carpets and drapes)
(Washable, reusable cat hair pickup roller.)
www.natlallergy.com
E-mail: info@nationalallergy.com
National Allergy Supply Inc.
1620-D Satellite Blvd.
Duluth, GA 30097
(800) 522-1448 (toll-free)

Defensive Decorating

Books

Szabo, Julia. *Animal House Style: Designing a Home to Share with Your Pets.* New York: Bulfinch Press, 2005.

Walker, Bob, and Frances Mooney. *The Cat's House.* Kansas City, MO: Andrews McMeel Publishing, 1996.

Walker, Bob, and Frances Mooney. *Cats into Everything.* Kansas City, MO: Andrews McMeel Publishing, 1999.

SureFit

(Slipcovers, pet throws, bed protectors.)
www.surefit.net
8000 Quarry Rd.
Alburtuis, PA 18011
(888) 796-0500 (toll-free)

Slipcover Shop

www.slipcovershop.com
58–25 Laurel Hill Blvd.
Woodside, NY 11377
(888) 405-4758 (toll-free; customer service weekdays 9 a.m. to 5 p.m. Eastern)
(718) 478-5234 (product information)

Waterproof Fabric Pad

(This is great! It protects mattresses and upholstered furniture against liquid with an extra-soft layer of vinyl bonded between two layers of polyester.)
The Vermont Country Store (Weston and Rockingham, VT)
http://vermontcountrystore.com
(800) 547-7849 (toll-free)

StickyPaws

(Sticky strips and other products to deter clawing.)
www.stickypaws.com
E-mail: orders@pioneerpet.com
Fe-Lines, Inc.

Pioneer Pet Products, LLC
N144W5660 Pioneer Rd.
Cedarburg, WI 53012
(866) 317-6278 (toll-free)

Zenith Tibet Almond Stick

(Hides scratches in wood furniture like magic.)
www.zcwinc.com
E-mail: zenithchemicalworks@zcwinc.com
Zenith Chemical Works, Inc.
133 Interstate Rd.
Addison, IL 60101
(630) 543-5161
Widely available at hardware stores and home centers

The Museum Putty, The Museum Wax, and The Museum Gel for Glass and Crystal

(For securing breakable items to shelves and tables.)
www.museum3pack.com
www.quakehold.com
E-mail: customerservice@readyamerica.com
Ready America, Inc.
Escondido, CA 92029
(800) 959-4053 (toll-free)
Widely available at Home Depot, Lowe's, OSH, ACE, TrueValue, and many others

Garden Defense

Coleus canina x Plectranthus ("scaredy-cat" coleus)

(Cat-repellent plant; cats hate the scent.)
www.gardeners.com
Gardener's Supply
128 Intervale Rd.
Burlington, VT 05401
(888) 833-1412 (toll-free)
(802) 660-3505; (802) 658-2433 (retail stores in Vermont)

Safety Products

100 Candles
(Rechargeable LED candles.)
www.100candles.com/rechargeable_candles.htm
E-mail: candles@100candles.com
100Candles.com
6263 Randolph St.
Commerce, CA 90040
(866) 611-8686 (toll-free)

Lamplust flame-free rechargeable lamps
(Cordless electric flame-free candle substitutes.)
www.lamplust.com
E-mail: info@lamplust.com
(212) 863-9736

Fire Shield
(Pet-safe extension cord; shuts off power if cat chews through cord.)
www.fireshield.com
www.trci.net
E-mail: productinfo@trci.net
Technology Research Corp.
5250 140th Ave.
North Clearwater, FL 33760
(800) 780-4324 (toll-free)

Cat Fence-In Cat Containment System
(Safe netting barrier that installs easily along the top perimeter of any wood, masonry, wire, or chain-link fence to keep a cat safely in the yard, and stray cats out.)
www.catfencein.com
PO Box 795, Department E
Sparks, NV 89432
(888) 738-9099 (toll-free)

Purrfect Fence
(Safe netting barrier that can be free-standing or added to an existing wall.)
www.purrfect.fence.com
E-mail: info@purrfectfence.com
Purr...fect Fence LLC
579-C Overton Place
Long Branch, NJ 07740
(888) 280-4066 (toll-free)

Affordable Cat Fence
(Safe netting barrier supported by heavy-duty steel stanchions.)
http://catfence.com
E-mail: info@catfence.com
722 W. Kings Hwy.
San Antonio, TX 78212
(888) 840-CATS (888-840-2287) (toll-free)

Computer Protection

Kitty Keyboard Kover
(Acrylic keyboard cover.)
www.kittykeyboardkover.com
E-mail: info@kittykeyboardkover.com

Kit-In Box
(Cat bed that attaches to your desk.)
www.therefinedfeline.com/kitinbox-cat-perch.htm
Facebook: The Refined Feline
The Refined Feline
(800) 289-6136 (toll-free)

PawSense
(Software program that detects "cat-like typing.")
www.pawsense.com
E-mail: bitboost.sales.x201@bitboost.com

BitBoost Systems
PO Box 65502
Tucson, AZ 85728
(520) 623-6326

"FurVeillance" Technology

The Eyenimal Video Camera for pets
(Video camera worn on a cat's collar.)
www.eyenimal.com/en
E-mail: contact@eyenimal.com
Available at www.spiritessences.com/eyenimal-pet-video-cam; www.dogtek.com

Mr. Lee Catcam
(Video camera [two hours of continuous video] or still photo camera [five hundred photos] worn on a cat's collar.)
www.mr-lee-catcam.de
Also available as a kit for electronics do-it-yourselfers

Panasonic Network Camera and Pet Cam BL-C1A
(Motion-sensor wireless camera that allows remote controlling of the camera, and lets you view your cat from your web browser, PC, PDA, cell phone, or other device.)
Available at amazon.com and other retailers

Behavior Modification Tools

Books

Krieger, Marilyn. *Naughty No More.* Irvine, CA: Bow Tie Press, 2011.
Pryor, Karen. *Getting Started: Clicker Training for Cats.* Waltham, MA: Sunshine Books, 2001.

Clicker Training
www.clickertraining.com
Karen Pryor Clicker Training (KPCT)

49 River Street, #3
Waltham, MA 02453
(800) 47-CLICK (toll-free)

CAT MASSAGE

Books

Ballner, Maryjean. *Cat Massage: A Whiskers-to-Tail Guide to Your Cat's Ultimate Petting Experience.* New York: St. Martin's Press, 2000.

Videos

Your Cat Wants a Massage! by Maryjean Ballner
www.catmassage.com
E-mail: maryjean@catanddogmassage.com
Cat and Dog Massage Works
PO Box 8206
Santa Barbara, CA 93118
(877) MEOW-MEOW (toll-free)

FELINE SYNTHETIC FACIAL PHEROMONES

Comfort Zone with Feliway Spray Stress-Reducing Pheromone for Kittens and Cats

www.petcomfortzone.com
Facebook: comfortzoneproducts
Central Garden & Pet
1340 Treat Blvd., Suite 600
Walnut Creek, CA 94597
(800) 234-2269 (toll-free)
Store locator: www.petcomfortzone.com/cats/product-locator.html

Comfort Zone with Feliway Diffuser Stress-Reducing Pheromone for Kittens and Cats

(Plug-in diffuser.)
Same contact as previous.

Grooming Products

The FURminator
(The best tool for detangling and de-shedding cat fur.)
www.furminator.com
Facebook: FURminator Inc.
info@FURminator.com
Blog: www.furminator.com/_blog/The_FURminator_Blog
(888) 283-1620 (toll-free)

KONG Zoom Groom for Cats
(Grooming and massaging tool—my cats love this.)
www.kongcompany.com/products/cats/e-collar-and-grooming/kong-zoom
groom-for-cats
E-mail: KONG@KONGCompany.com
Widely available at pet supply stores

Claw Caps
Soft Claws Nail Caps for Cats
www.softclaws.com
Dr. Christianne Schelling
39685 Sierra Dr.
Three Rivers, CA 93271
(800) 989-2542 (toll-free, 24/7)

Dental Care Products

Hills Prescription Diet t/d Feline Dental Health
(A complete, balanced dental health diet—we use it as cat treats.)
www.hillspet.com
Available through veterinarians.

C.E.T. Enzymatic Toothpaste for Dogs and Cats
(My cats love the poultry flavor!)
www.virbacvet.com/products/DentalHealth/ToothpastesAndToothbrushes/
C.E.T.PetToothpastes.aspx
Virbac Corporation

3200 Meacham Blvd.
Fort Worth, TX 76137-4611
(800) 338-3659 (toll-free)

Cat Scales

Digital Cat Scales

(For weighing your cat accurately at home.)
www.zieis.com/Cat-Scales-C168.aspx
ZIEIS
15050 Cedar Ave. S., #116-337
Apple Valley, MN 55124
(952) 461-2695 (support)
(877) 253-6017 (toll-free sales; Mon through Fri 10 a.m. to 4 p.m. Central)

Cat Treats

Wildside Salmon
(The BEST cat treat ever—freeze-dried pure wild Alaskan salmon bites.)
www.wildsidesalmon.com
E-mail: wildsidepetproducts@gmail.com
Wildside Pet Products
4411 S. Ferdinand St.
Seattle, WA 98118
(206) 722-FISH
Also available at www.catniptoys.com

Kitty Kaviar
(Freeze-dried flakes of shaved bonito fish—my kitties love them.)
www.kittykaviar.com
Available at www.catniptoys.com

Environmental Enrichment

INTERACTIVE TOYS

The Feline Fisher (formerly "Galkie Kitty Tease")
(THE best cat toy ever!)
www.felinefisher.com
E-mail: info@felinefisher.com
Red Cat Company
PO Box 57
Beaverton, OR 97075
(503) 626-2019

Da Bird
(The best interactive fishing-pole-type feather toy.)
www.go-cat.com
Go Cat Feather Toys
Harriet Morier, owner
3248 Mulliken Rd.
Charlotte, MI 48813
(517) 543-7519 (phone and fax)
Available online and at many pet supply stores

Whirly-Bird Cat Exercise Toy
(Fishing-pole-type interactive toy—cats love to bite down on the lure.)
www.whirly-bird.com
E-mail: info@whirly-bird.com
Aeries Enterprises
830 E. Pierson Dr.
Lynn Haven, FL 32444
(850) 265-2924

Fly Toy
(Irresistible "wiggling insect" fishing-pole toy.)
www.metpet.com
Metropolitan Pet
PO Box 230324
Tigard, OR 97281

(800) 966-1819 (toll-free)
Available at many pet stores.

Neko Fly
(Fishing-pole-type interactive toy with realistic insect-like lures.)
www.nekoflies.com
E-mail: customerservice@nekochan.ca
Nekochan Enterprises, Inc.
1685 H St., #421
Blaine, WA 98230
(866) 699-6356 (toll-free)

SOLO TOYS

Bat-a-Rat
(Feeding, foraging, and exercise toy.)
www.batarat.com
Facebook: Bat-A-Rat
Available online and at select pet stores

Enchantacat Toys
(Very sturdy solo toys filled with organic catnip.)
www.favoritepetproducts.com
Facebook: Plain Brown Tabby Toys and Treats
Available at www.catniptoys.com

Hot Cats
(Catnip-filled "cigars.")
Facebook: Plain Brown Tabby Toys and Treats
Available at www.catniptoys.com

Yeowww! Catnip Toys
(Potent organic catnip-filled toys in many cute and clever shapes.)
www.duckyworld.com
Facebook: Yeowww! Catnip
E-mail: info@duckyworld.com
(866) 493-6999 (toll-free)
Also available at www.catniptoys.com

Cosmic Kittyherbs (Kitty grass seed.)
Alpine Scratcher (Cardboard scratch mat.)
Cosmic Catnip (World's leading brand of catnip!)
(Plus lots of other great cat products.)
www.cosmiccatnip.com
Facebook: Cosmic Catnip
E-mail: info@cosmiccatnip.com
Cosmic Catnip, an OurPet's Brand
1300 East St.
Fairport Harbor, OH 44077
(440) 354-6500 (Mon through Fri, 8 a.m. to 5 p.m. Eastern)
Available at PetSmart, PetCo, and other pet supply retailers

Kitty Can't Cope Sack
(Small very sturdy cotton sack filled with 100 percent fresh catnip—my cats love these!)
www.kittycantcopesack.com
E-mail: use web form or kccs@maine.rr.com
Back Bay Pet Co.
PO Box 801
Portland, ME 04104
(207) 878-3826

Pawbreakers
("Catnip candy for cats"—combination toy and treat—1¼-inch-diameter ball of compressed catnip.)
www.edibleanimaltreats
E-mail: sales@edibleanimaltreats.com
E.A.T.s
330 Detroit Ave., Suite A
Monroe, MI 48162
(877) 228-8691 (toll-free)

Play-N-Squeak
(Microchip sound toy mouse and other toys.)
www.playnsqueak.com
Facebook: PlayNSqueakCatToys

E-mail: info@playnsqueak.com
Play-n-Squeak, an OurPet's Brand
1300 East St.
Fairport Harbor, OH 44077
(440) 354-6500 (Mon through Fri, 8 a.m. to 5 p.m. EST)

Kookamunga Catnip
(Very potent catnip.)
www.eightinonepet.com
E-mail: info@eightinonepet.com
Spectrum Brands, Inc.
Eight in One Pet Products
Islandia Pavilion, Suite 100
1377 Motor Pkwy.
Islandia, NY 11749
(800) 645-5154 (toll-free)
Available at PetSmart, PetCo, and other pet supply retailers

KITTY VIDEOS

DVD videos for cats
(DVDs of day and night views of live bugs and birds.)
www.kittyshow.com
E-mail: 2milo@KittyShow.com
Kitty Show
4032 Shell Point Rd.
Beaufort, SC 29906
(843) 271-0747

Video Catnip
(*The* original video for cats; other videos made for cats are also available.)
www.cattv.com
E-mail: info@cattv.com
Pet AVision, Inc.
245 Morris St.
Morgantown, WV 26501-7524
(304) 292-1272

Scratching Posts, Climbing Trees, Cat Furniture

Angelical Cat Company
(Investment-quality cat furniture, climbing trees, and scratching posts.)
www.angelicalcat.com
E-mail: generalinfo@angelicalcat.com
9311 NW 26 Place
Sunrise, FL 33322
(954) 747-3629

Ace Cat Furniture
(Wide range of cat trees, scratching posts, and other feline furniture.)
www.acecatfurniture.com
Facebook: Ace Cat Furniture
E-mail: sales@acecatfurniture.com
202 Blue Creek Dr.
Urbana, IA 52345
(800) 404-7068 (toll-free)

Drs. Foster & Smith, Inc.
(A good selection of sturdy cat furniture and lots more.)
www.drsfostersmith.com
Facebook: Drs. Foster and Smith Pet Supplies
E-mail: customerservice@drsfostersmith.com or use web form
2253 Air Park Rd., PO Box 100
Rhinelander, WI 54501
(800) 381-7179 (toll-free)

Cat Exercise Wheel
(Like a hamster wheel—but for cats!)
http://catwheelcompany.com
E-mail: info@catwheelcompany.com or use web form
10 Hughes St., Suite A-202
Irvine, CA 92618
(949) 529-5531
Supplier locator: www.catwheelcompany.com/usadistributors.cfm
Also available at www.oldmaidcatlady.com

KittyWalk systems
(Cat strollers, outdoor tunnels and enclosures, gazebos, climbing structures for indoor cats, carriers, and beds.)
www.kittywalksystems.com
E-mail: kittywalk5@optonline.net
(877) 548-8905 (toll-free)

Electric Water Fountains

Drinkwell Pet Fountain
(Our favorite fountain: the Drinkwell 360. Easy to clean, and the kitties love it!)
www.petfountain.com
E-mail: info@petfountain.com
Veterinary Ventures, Inc.
1135 S. Rock Blvd., Suite #330
Reno, NV 89502
(866) 322-2530 (toll-free)
Available widely at pet supply retailers and at www.petsafe.net.

Petmate Fresh Flow Pet Fountain
www.petmate.com
E-mail: use web form
Petmate
PO Box 1246
Arlington, TX 76004-1246
Also available at PetSmart, PetCo, and others.

One-stop Shopping for Cat Goodies

Mr. Chewy
(Supplies, food, toys, accessories, etc.)
www.mrchewy.com
Facebook: MrChewyPets

Old Maid Cat Lady
(All things cat for cats and their people.)
http://oldmaidcatlady.com

Resources for Strays, Ferals, Cat Colonies, Spay/Neuter, TNR, and Fostering

Books

Berkeley, Ellen Perry. *TNR: Past, Present and Future: A History of the Trap-Neuter-Return Movement.* Washington, DC: Alley Cat Allies, 2004.

Boneham, Sheila Webster, PhD. *Rescue Matters: How to Find, Foster, and Rehome Companion Animals: A Guide for Volunteers.* Crawford, CO: Alpine Publications, 2009.

Holton, Louise. *Feral Cat Colony Management and Control: Facts and Myths about Feral Cats.* Mount Rainier, MD: Alley Cat Rescue, 2002. (Available at www.saveacat.org/catfacts.html.)

Kreuz, Tamara. *The Stray Cat Handbook.* Foster City, CA: Howell Book House, 1999.

Community Resources

Alley Cat Allies
(Extensive help and resources for spay/neuter and TNR.)
"The National Feral Cat Resource"
www.alleycat.org
Facebook: Alley Cat Allies
E-mail: info@alleycat.org
7920 Norfolk Ave., Suite 600
Bethesda, MD 20814-2525
(240) 482-1980

Alley Cat Rescue, Inc.
(Extensive help and resources for spay/neuter and TNR.)
"National Cats Protection Association"
www.saveacat.org
Facebook: Alley Cat Rescue Inc.
Blog: http://alleycatrescue.blogspot.com
E-mail: acr@saveacat.org
Louise Holton, president and founder
PO Box 585
Mount Rainier, MD 20712
(301) 277-5595

Feral Cat Coalition
(Help and resources for spay/neuter and TNR.)
www.feralcat.com
E-mail: use web form
9528 Miramar Road, PMB 160
San Diego, CA 92126
(619) 758-9194

Stray Cat Alliance
(Help with spay/neuter and TNR: "Building a no-kill nation one stray at a time.")
www.straycatalliance.org
Facebook: StrayCatAlliance
E-mail: info@straycatalliance.org
PO Box 661277
Los Angeles, CA 90066

Feral Friends Community Cat Alliance
(Help and mentoring with spay/neuter and TNR.)
www.feralfriends.org
E-mail: info@feralfriends.org
PO Box 832857
Richardson, TX 75083-2857
(972) 671-0429

Feral Cat Caretakers' Coalition
(Helpful articles about TNR, fostering, and adopting and caring for ferals.)
www.feralcatcaretakers.org
E-mail: info@feralcatcaretakers.org
PO Box 491244
Los Angeles, CA 90049
(310) 820-4122

Neighborhood Cats
(Resources and tools for TNR, rescue, and fostering of feral cats.)
www.neighborhoodcats.org
Facebook: Neighborhood Cats
E-mail: headcat@neighborhoodcats.org

2576 Broadway, No. 555
New York, NY 10025
(212) 662-5761

FixNation

(Los Angeles–based free spay/neuter clinic; website has lots of resources for feral
caretakers.)
www.fixnation.org
Facebook: FixNationLA
E-mail: info@fixnation.org
7680 Clybourn Ave.
Los Angeles, CA 91352
(818) 524-2287

Sources of Reliable Information about Cats

Book

Cat Owner's Home Veterinary Handbook. Fully revised and updated by Debra M.
Eldredge, DVM, et al. Foster City, CA: Howell Book House, 2007.

American Association of Feline Practitioners (AAFP)

www.catvets.com

CATalyst Council

www.catalystcouncil.org
Free downloadable copy of *CATegorical Care: An Owner's Guide to America's #1
Companion*
www.catalystcouncil.org/resources/health_welfare/categorical_care

The Cat Fanciers' Association

(The world's largest registry of pedigreed cats—lots of information on cat breeds,
health, care, and much more.)
www.cfa.org

Cornell Feline Health Center

www.vet.cornell.edu/fhc

The Indoor Cat Initiative at the Ohio State University College of Veterinary Medicine
indoorpet.osu.edu/index.cfm

TICA—The International Cat Association
(The world's largest genetic registry of pedigreed cats—lots of information on cat breeds, health, care, and much more.)
www.tica.org

Winn Feline Foundation
www.winnfelinehealth.org

To Find a Certified Cat Behavior Consultant

IAABC—The International Association of Animal Behavior Consultants
www.iaabc.org

The CFA's Cat Museum

The Cat Fanciers' Association Historical Foundation
www.felinehistoricalfoundation.org
Facebook: CFA Foundation Inc.
Museum located at
260 E. Main St.
Alliance, OH 44601
(330) 680-4444

Index